NEUROSCIENCE
INTELLIGENCE
UNIT

Cell-Cycle Mechanisms and Neuronal Cell Death

Agata Copani, M.D., Ph.D.
University of Catania
Department of Pharmaceutical Sciences
Catania, Italy

Ferdinando Nicoletti, M.D.
Department of Human Physiology and Pharmacology
University of Rome "La Sapienza"
Rome, Italy
and
I.N.M. Neuromed
Pozzilli, Italy

LANDES BIOSCIENCE / EUREKAH.COM
GEORGETOWN, TEXAS
U.S.A.

KLUWER ACADEMIC / PLENUM PUBLISHERS
NEW YORK, NEW YORK
U.S.A.

CELL CYCLE MECHANISMS AND NEURONAL CELL DEATH

Neuroscience Intelligence Unit

Landes Bioscience / Eurekah.com
Kluwer Academic / Plenum Publishers

Printed in the U.S.A.

Kluwer Academic / Plenum Publishers, 233 Spring Street, New York, New York, U.S.A. 10013
http://www.wkap.nl/

Please address all inquiries to the Publishers:
Landes Bioscience / Eurekah.com, 810 South Church Street, Georgetown, Texas, U.S.A. 78626
Phone: 512/ 863 7762; FAX: 512/ 863 0081
http://www.eurekah.com
http://www.landesbioscience.com

Cell-Cycle Mechanisms and Neuronal Cell Death, edited by Agata Copani and Ferdinando Nicoletti, Landes / Kluwer dual imprint / Landes series: Neuroscience Intelligence Unit

ISBN: 0-306-47850-1

While the authors, editors and publisher believe that drug selection and dosage and the specifications and usage of equipment and devices, as set forth in this book, are in accord with current recommendations and practice at the time of publication, they make no warranty, expressed or implied, with respect to material described in this book. In view of the ongoing research, equipment development, changes in governmental regulations and the rapid accumulation of information relating to the biomedical sciences, the reader is urged to carefully review and evaluate the information provided herein.

Library of Congress Cataloging-in-Publication Data

Cell-cycle mechanisms and neuronal cell death / Agata Copani, Ferdinando Nicoletti.
    p. ; cm. -- (Neuroscience intelligence unit)
  Includes bibliographical references and index.
  ISBN 0-306-47850-1
  1. Nervous system--Degeneration--Pathophysiology. 2. Neurons. 3. Apoptosis. 4. Cell cycle. I. Copani, Agata. II. Nicoletti, F. (Ferdinando) III. Series: Neuroscience intelligence unit (Unnumbered)
  [DNLM: 1. Apoptosis. 2. Cell Cycle. 3. Neurons--cytology. 4. Neurodegenerative Diseases--etiology. WL 102.5 C39245 2005]
  RC365.C45 2005
  616.8--dc22

                              2005013874

CONTENTS

EDITOR

Agata Copani
Department of Pharmaceutical Sciences
University of Catania
Catania, Italy
Email: acopani@katamail.com
Chapter 7

Ferdinando Nicoletti
Department of Human Physiology and Pharmacology
University of Rome "La Sapienza"
Rome, Italy
and
I.N.M. Neuromed
Pozzilli, Italy
Email: ferdinandonicoletti@hotmail.com
Chapter 7

CONTRIBUTORS

Thomas Arendt
Paul Flechsig Institute of Brain Research
Department of Neuroanatomy
University of Leipzig
Leipzig, Germany
Email: aret@medizin.uni-leipzig.de
Chapter 1

Craig S. Atwood
Institute of Pathology
Case Western Reserve University
Cleveland, Ohio, U.S.A.
Chapter 6

Filippo Caraci
Department of Pharmaceutical Sciences
University of Catania
Catania, Italy
Chapter 7

Andrea Caricasole
SienaBiotech
Siena, Italy
Chapter 7

Karl Herrup
Alzheimer Research Lab
University Hospitals of Cleveland
Case Western Reserve University
School of Medicine
Cleveland, Ohio, U.S.A.
Email: kxh26@cwru.edu
Chapter 2

Mervyn J. Monteiro
Laboratory of Neurodegenerative Studies
Medical Biotechnology Center
Baltimore, Maryland
Email: monteiro@umbi.umd.edu
Chapter 3

Laura Morelli
Institute of Pathology
Case Western Reserve University
Cleveland, Ohio, U.S.A
Chapter 6

Zsuzsanna Nagy
Neuroscience Division
Medical School
University of Birmingham
Birmingham, U.K.
Email: z.nagy@bham.ac.uk
Chapter 4

Mark E. Obrenovich
Institute of Pathology
Case Western Reserve University
Cleveland, Ohio, U.S.A.
Chapter 6

Osamu Ogawa
Institute of Pathology
Case Western Reserve University
Cleveland, Ohio, U.S.A.
Chapter 6

Huntington Potter
Johnnie B. Byrd Sr. Alzheimer's Center
   and Research Institute
and
Suncoast Gerontology Center
and
Department of Biochemistry
   and Molecular Biology
University of South Florida College
   of Medicine
Tampa, Florida, U.S.A.
Email: hpotter@byrdinstitute.org
Chapter 5

Arun K. Raina
Institute of Pathology
Case Western Reserve University
Cleveland, Ohio, U.S.A.
Chapter 6

Mark A. Smith
Institute of Pathology
Case Western Reserve University
Cleveland, Ohio, U.S.A.
Email: mas21@po.cwru.edu
Chapter 6

Maria Angela Sortino
Department of Experimental
   and Clinical Pharmacology
University of Catania
Catania, Italy
Chapter 7

Yan Yang
Alzheimer Research Lab
University Hospitals of Cleveland
Case Western Reserve University
School of Medicine
Cleveland, Ohio, U.S.A.
Email: yxy33@cwru.edu
Chapter 2

PREFACE

The aim of this book is to provide an overview of current evidence relating the process of neuronal death to the aberrant reentry of differentiated neurons into the cell cycle. The seven chapters of the book trace the development of the new concept that mitotic reactivation is instrumental to neuronal death in many neurodegenerative conditions, including Alzheimer's disease, stroke, amyotrophic lateral sclerosis, and epilepsy. Of significance to the pathogenesis of these diseases is the presence of cell division markers in selectively vulnerable neurons. Protein-based evidence for the reactivation of a cell cycle in neurons has been strengthened by the demonstration of *bona fide* DNA replication in at-risk neurons in Alzheimer's brain. Although this pattern is evocative of a real cell cycle, there is no evidence for neuronal mitosis in diseased brains, but rather a persistent association between cell cycling and neuronal loss. Tissue culture studies have shown that blocking the cell cycle could prevent neuronal death. Thus, forcing the cycling of an adult neuron might result in cell death.

In this book we have collected some current views of the link between cell cycle and cell death in neurons. Arendt hypothesizes that cell cycle changes underlie a brain-morpho-dysregulation. Copani and colleagues discuss the biochemical details of the neuronal cell cycle. Monteiro implicates presenilins in cell cycle regulation and death. Nagy summarizes evidence for cell cycle-related events as a convergent mechanism in different neurodegenerative diseases. Obrenovich and colleagues recount some of the mitotic alterations, including the recruitment of mitogenic factors and oxidative stress, in Alzheimer's disease. Potter expounds the idea that defects in mitosis and particularly in chromosome segregation may be a part of the Alzheimer's disease process. Yang and Herrup give an overview of the participation of cell cycle events in human diseases and in mouse models of Alzheimer's disease. The book envisages that the discovery of links between the cell cycle and cell death might hold the key to early diagnosis and treatments of neurodegenerative diseases, providing the background for an in-depth and thorough examination of the discovery.

Agata Copani
Ferdinando Nicoletti

CHAPTER 1

Cell Cycle Activation in Neurons:

The Final Exit of Brain-Morpho-Dysregulation

Thomas Arendt

Mammalian cells act and react within frameworks of defined cellular programmes such as division, movement, adhesion, differentiation and death, integrating both genetic and epigenetic information. Execution of these programmes, ones activated, are surprisingly stable, even under pathological conditions. Disorders of mammalian organisms are, therefore, neither due to "disease programmes" nor to complete dysruptions of the orderly execution of normal cellular programmes rather than to an activation or termination of these physiological programmes wrongly placed in time and space. This principle applies to all three basic cellular pathological reactions such as inflammation, neoplastic transformation or degeneration. Therefore, any endogenous or exogenous pathological stimulus can give rise to only a very limited spectrum of cellular reactions.

Inflammation, neoplastic transformation and degeneration to a large extend share cellular programmes and make use of identical signaling networks. Control of proliferation and differentiation, i.e., cell cycle control, is not only the major issue for both inflammation and neoplastic transformation, it also is critically involved in degeneration of nondividing cells, such as mature neurons,[1,2] Errors in cell cycle regulation can, thus, lead to uncontrolled cell growth and division such as in autoimmune diseases, cancer or neurodegeneration.

Understanding AD and neurodegeneration within the framework of cellular differentiation and cell division control is not a new idea. It had first been brought up early in the 20th century,[3] and repeatedly been stressed in the past as concept of 'hyperdifferentiation',[4] 'retrodifferentiation'[5] 'dysdifferentiation'[6] or 'dedifferentiation'.[1,2] In the early ninetieth of the last century, it became increasingly clear that mitogenic pathways in neurons are aberrantly activated early during AD.[7] From that, further insigths into the pathomechanism progressed further downstream, suggesting a potential involvement of cell cycle control as a result of abnormal mitogenic signaling more recently.[8-10]

Selective Neuronal Vulnerability: The Key Issue to Understand Neurodegenerative Disorders

Cell death in the nervous system, irrespectively as to whether caused by inflammation, ischemia or primary degeneration selectively affects special types of neurons while others are spared rather constantly. In AD, this selective neuronal vulnerability has systemic character. AD, thus, is a truely "neuromorphological disorder", as "it knows where to go when".

Understanding this selective neuronal vulnerability means understanding Alzheimer's disease. More than one hundred years ago, Paul Flechsig[11] had advanced the idea that variations in vulnerability of different groups of neurons are to be traced back in large part to developmental conditions, a concept, later defined by Cecile and Oscar Vogt (1951)[12] as the "principle of pathoclisis". Brain areas and neuronal types highly vulnerable against neurofibrillary degeneration in AD are indeed different to the rest of the brain at least with respect to three other aspects which might be the key to understand underlaying mechanisms: they have been acquired late during phylogenetic development (or have completely been reorganized lately), they mature rather late during ontogenic development and they exhibit a particularly high degree of synaptic plasticity (and probably synaptic turnover).

Neurons are the most highly differentiated cells in the true sense of the word. After proliferation, migration and differentiation is completed, they become integrated for a lifetime into a neuronal network. This integration requires intercellular communication which is largely regulated through external cues. Connectivity and attachement of a cell are mechanisms that during evolution from single cellular systems to multicellular systems have been acquired to interpret signals from the environment and respond appropriately by proliferation, differentiation and cell death. In a multicellular organism, external cues are, thus, used for morphoregulation, i.e., the assembly of individual cells into highly ordered tissues. Differentiated, nonproliferative neurons, however, make use of these mechanisms that control connectivity for their genuine function i.e., for information processing in a multicellular network. Cell-cell-interactions in the nervous system subserve a newly acquired purpose, the formation of a network. Once the network had been formed, the cell cycle is shut down and information from neighbouring cells is translated into continous changes of synaptic strength and morphology, a process, referred to as synaptic plasticity. This means, at least partially identical mechanisms are used to regulate such divergent effectors as cell cycle control and synaptic plasticity.

For the neuron, it might be a great achievement to control its synaptic plasticity, to do this on the potential expense of differentiation control, however, might put the neuron on a permanent risk. Neurons acquired particularly late during evolution of the human brain (e.g., cortical associative circuits) that subserve 'typical human higher cortical functions' such as learning, memory, reasoning, consciousness etc. need to display a particularly high degree of synaptic plasticity which might explain the particularly high sensitivity towards the loss of differentiation control and cell death (Fig. 1).

Plasticity Meets Cell Death

Alterations of neuroplasticity are a constant element of the pathomechanism of AD.[13-30] It is the plastic morphogenetic potential that "creates" the brain, tightly linked to both synaptic pasticity and cell cycle control, what is impaired in AD.

Synaptic plasticity is thought to be the major mechanism underlaying information processing and memory formation in the brain. There is evidence, moreover, that links it to various clinical syndromes, including amnesia and dementia, epilepsy[31] and probably psychosis.[32] Such seemingly diverse phenomena such as beneficial alteration as learning and memory on one hand and pathological changes on the other, may possess common characteristics and molecular mechanisms providing the basis for a plasticity-pathology-continuum.[33] What makes the human brain so special, a large association cortex characterized by a particulary high degree of plasticity—a prerequisite for self conscioussness i.e., the development of a belief of itself—at the same time creates a vulnerability unique to human (Fig. 2). Identical signaling pathways might be used in the first stage of synaptic change in both development and maturity, and higher cognitive functions such as learning and memory might be based on adaptive modifications of an ancient mechanism initially evolved to wire the brain. The activity-dependent modification of synapses is, thus, a powerful mechanism for shaping and modifying the response properties of neurons, but it is also dangerous[34] as it still hides its 'beasty potency' of phylogenetic regression into cell cycle activation which ultimately results in cell death.[35]

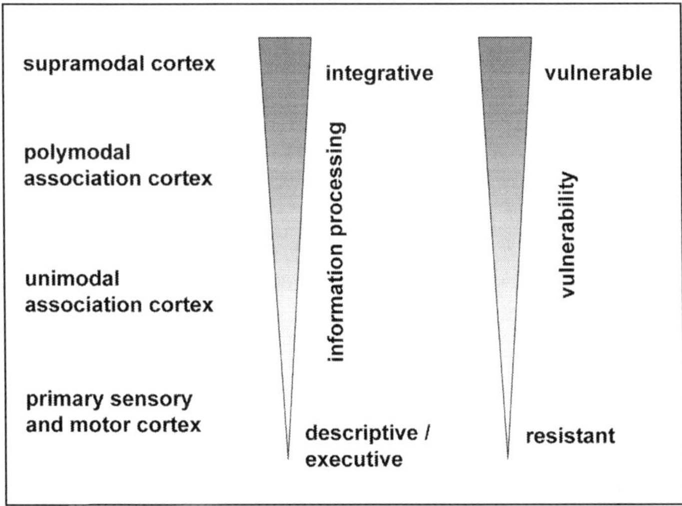

Figure 1. Hierarchy of cortical vulnerability against neurofibrillary degeneration compared to the neuroanatomical hierarchy of cortical information processing. The progression of neurofibrillary degeneration throughout different cortical areas in AD follows a sequence that represents systematic differences in vulnerability which match the hierarchic pattern of cortical information processing.

Dynamic Morphodysregulation

Neurodegeneration in Alzheimer's disease is associated with aberrant neuritic growth.[36] Abnormal growth profiles preferentially affect neurons that are potentially prone to neurodegeneration such as cholinergic basal forebrain neurons and cortical pyramidal cells (Fig. 3). Aberrant sprouts can be detected in early stages of the disease, precede the formation

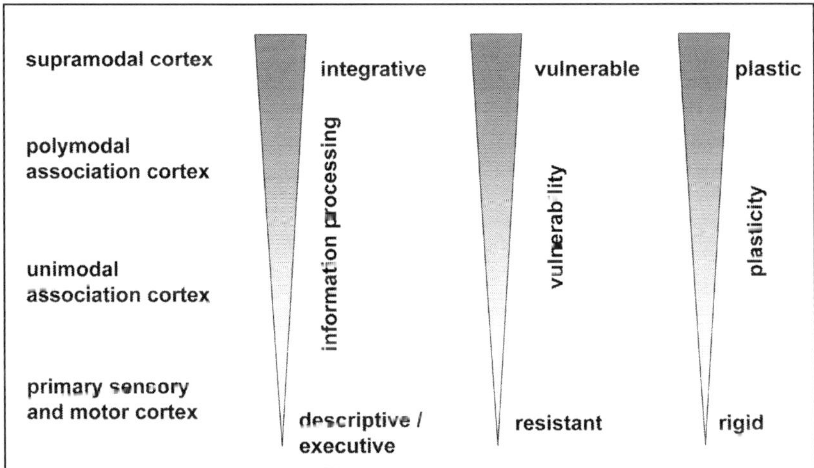

Figure 2. Hierarchy of plastic capacity of pyramidal neurons in the cerebral cortex as compared to the hierarchy of cortical vulnerability against neurofibrillary degeneration in the cerebral cortex and to the neuroanatomical hierarchy of cortical information processing. The degree of plasticity is highest in supramodal and polymodal association cortices, less well developed in unimodal association cortices and only marginally present in primary sensory and motor cortices.

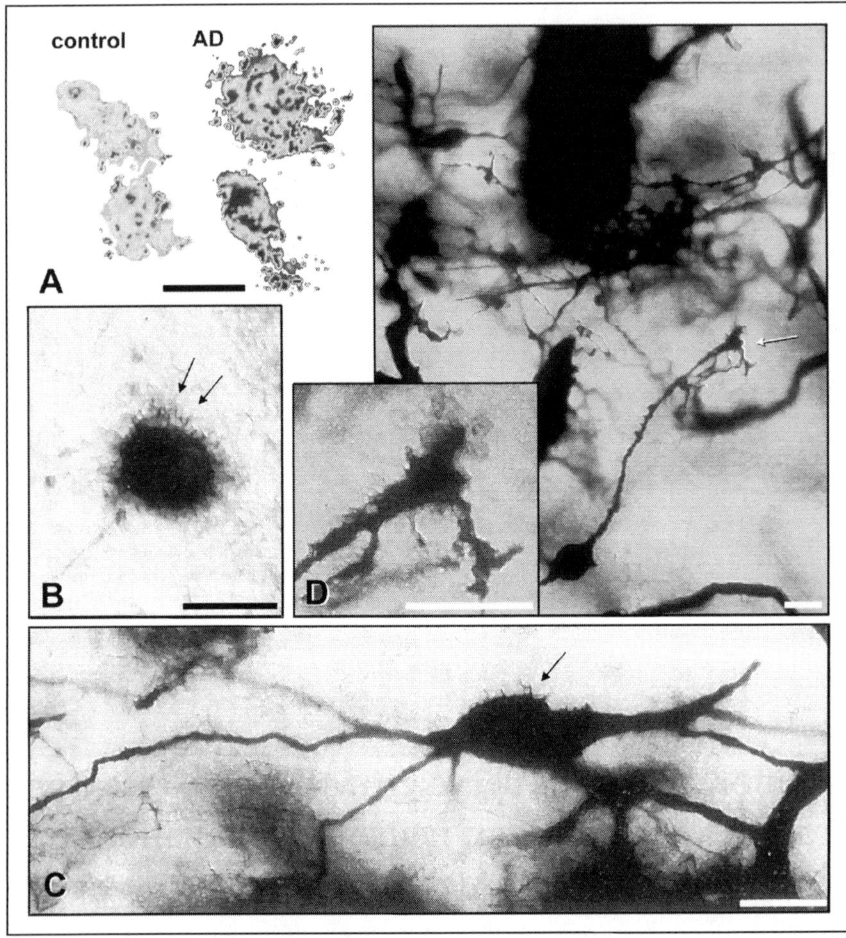

Figure 3. Abberant morphological growth profiles in AD on neurons of the cholinergic basal nucleus of Meynert. Growth profiles are localized on dendritic endings (D) or cover the soma (B, C). The appearance of these growth profiles is associated with an upregulation of the NGF receptors TrkA (A) and p75 (B). (A= TrkA insitu hybridisation; B= p75 immunocytochemistry; C, D= Golgi impregnation.) Scale bar= 30 μm. (Figure modified from refs. 37,39.)

of paired helical filaments and occur even without massive neuronal loss.[37] They might, thus, represent an event of primary significance, inherent to the pathomechanism rather than a secondary event triggered by ongoing degeneration.

As opposed to the continous growth of dendritic elements during aging,[38] growth in AD is aberrant with respect to its localisation, morphological appearance, and composition of cytoskeletal elements[36,39-42] (Fig. 4). Dystrophic neurites, mainly dendritic but occasionally also axonal in origin form a constant component of AD pathology. The original identification of these neurites as aberrant sprouts[43] about 100 years ago, has been corroborated more recently by Golgi studies, ultrastructural evidence and the accumulation of growth-associated proteins(for rev. see Ref. 35). Growth profiles on cholinergic basal forebrain neurons, for example, are typically associated with receptors for NGF indicating the potential ability to respond to an increased trophic force[37] (Fig. 3).

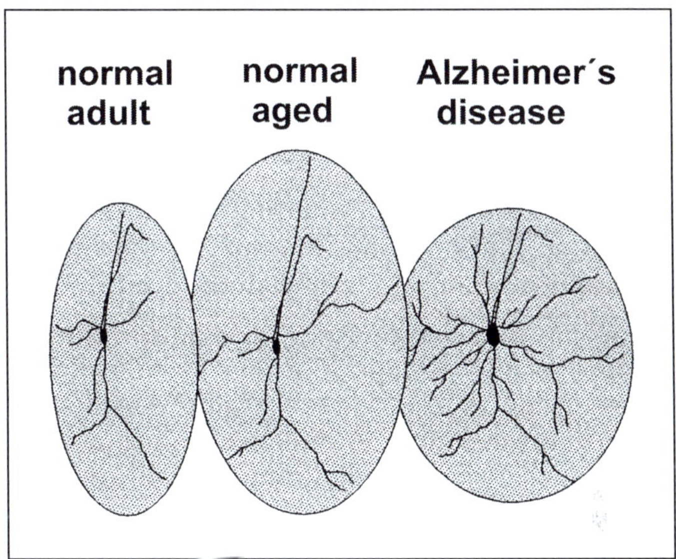

Figure 4. Synopsis of principal changes in the dendritic organization of neurons in the cholinergic basal nucleus of Meynert during aging and in AD. While during aging, dendritic growth preferentially occurs on dendritic endings, thereby contributing to a significant extension of the overall dendritic area ('extensive' dendritic growth), dendritic growth in AD is confined to more proximal parts of the dendritic tree, resulting in more dense dendritic ramification with only marginal effects on the extension of the overall dendritic area ('intensive' dendritic growth). (Figure modified from ref. 39.)

Elevated Mitogenic Force

The presence in the AD brain of growth-associated and growth-promoting proteins, as well as their receptors might be an indication of an increased mitogenic force particularly pronounced within the microenvironment of plaques. The still growing list of compounds contains GAP-43, MARCKS, spectrin, heparansulfate, laminin, NCAM, various cytokines and neurotrophic factors such as NGF, bFGF, EGF, IL-1, IL-2, IL-6, IGF-1, IGF-2, PDGF, vascular endothelial growth factor and HGF/SF (for review see ref. 35).

Activated Mitogenic Signaling

p21ras

A number of these neurotrophic and potentially mitogenic compounds which are elevated early in the course of the disease mediate their cellular effects through activation of receptor tyrosine kinases, which recruit p21ras, a GTPase that belongs to the Ras superfamily of small-G-proteins and lead to the sequential activation of Raf, MEK and ERK.[44] The sequential activation of the MAPK-cascade allows for multiple interactions through phosphorylation and dephosphorylation which will modify extracellular signals. The Ras-ERK/MAPK pathway is an evolutionary conserved signaling mechanism that plays a fundamental role in regulation of cellular proliferation and differentiation and control of neuroplasticity and is also involved in modulating the expression and posttranslational processing of APP and tau protein.[45-47]

In mammalian cells, the p21ras gene product is encoded by a family of ras proto-oncogenes that include at least three functional loci, H-ras, K-ras and N-ras. They are activated when bound to GTP and inactivated upon hydrolyzing the gamma phosphate to form GDP.[48] Ras proteins function as 'molecular switches' that cycle between an inactive GDP-bound and an active GTP-bound state. Activation occurs by interaction with GDP-GTP exchange factors (GEFs). Binding of the neurotrophins to tyrosine kinase receptors (trk) converts p21ras from its inactive, GDP-bound, to active, GTP-bound, state. GTP-bound p21ras recruits raf-kinase from the cytoplasm to the plasma membrane, where it is activated.[49] Raf-kinase phosphorylates and activates the mitogen activated protein kinase kinase (MAPKK) leading to the activation of the mitogen activated protein kinase (MAPK).

Ras proteins are involved in the central coordination of a variety of functions including cytoskeletal organization, gene expression, cell cycle progression, membrane trafficking, cell adhesion, migration and polarity, and synaptic plasticity.[50,74] During brain development, p21ras participates in the regulation of the G_0/G_1 transition of the cell cycle and might, thus, be a critical regulator for cellular proliferation and differentiation.[51] In the adult nervous system, it plays a role in reactive dendritic proliferation and neosynaptogenesis.[52-54] Ras, thus, is a central regulator of synaptic plasticity in the adult brain.[50]

In AD, p21ras is highly expressed in vulnerable brain areas already prior to its affection by neurofibrillary degeneration[55] which indicates an activation of the ras-MAPK pathway in very early stages of the disease.[56]

MAPK

The Mitogen-Activated-Protein-Kinases (MAPKs) or Extracellular Signal Regulated Kinases (ERKs) and MAPK-kinase or MAP/ERK kinase (MEK) belong to a group of protein kinases which is highly conserved from yeast to vertebrates.[57] They are key molecules in signal processing that become activated in response to a wide variety of reagents. Among these are tumour promotors, interleukins, growth factors whose receptors are tyrosine kinases, mitogens whose receptors couple to heterotrimeric guanine nucleotide binding proteins (G proteins), and agents that induce N-methyl-D-aspartate receptor activation. When activated, ERKs rapidly phosphorylate targets that lead to changes in kinase cascades, protein function, or gene expression. Effectors include Ser/Thr kinases (pp90[rsk], MAPK-activated protein kinase-2 and 3p-kinase), transcription factors (Elk-1, c-Myc, c-Jun, NF-116, ATF-2 and CREB) and structural proteins (talin, microtubule-associated proteins and lamins (for review see ref. 58).

The activation of MAPK that is associated by its nuclear translocation plays an essential role in the expression of many immediate-early and late response genes. MAPK activation and nuclear translocation are required for long-term facilitation in Aplysia[59] and LTP in vertebrates.[60] MAPK signaling, might thus be a critical regulator for both short-term synaptic function and transcription of genes required for long-term plasticity. Ras/MAPK signaling components are highly enriched in the adult CNS, and expression of many MAPK regulators is largely restricted to the CNS where they are highly abundant in association areas implicating a role in memory consolidation and synaptic plasticity.

ERK1/2 and other members of the MAPK family[61,62] are able to phosphorylate the microtubule-associated protein tau on threonine and serine residues found phosphorylated in PHFs. In AD, we observed indications for increased expression and activation of both MAPKK and MAPK already in very early stages of the disease.[56] This activation of the MAPK pathway can be modeled in transgenic mice expressing permantenly activated ras (Fig. 5).

Nitric Oxide and the Autocrine Loop of Selfperpetuating Ras/MAPK Activation

nNOS was originally thought to be a constitutively expressed enzyme. It becomes increasingly clear now, however, that its levels are dynamically regulated in response to neuronal development, plasticity and injury.[63] The transcriptional induction of nNOS that is controlled by neurotrophins and other growth factors is in turn involved in regulating the expression of immediate early genes in neurons thereby controlling neuronal growth and differentiation.[64,65]

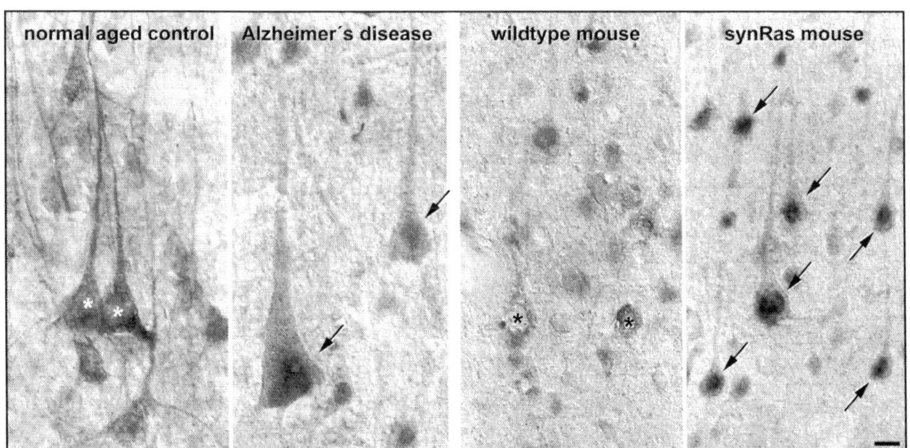

Figure 5. Changes in the subcellular distribution of MAP kinase (ERK1/2) in AD, and in a transgenic mice model expressing permanently activated p21ras under control of the synapsin promotor (synRas mouse). Both in AD and transgenic mice, MAPK is translocated from the cytoplasm (asterisk) to the nucleus (arrows) indicating an activation of the enzyme. Scale bar: 10 μm (modified after refs. 53,56).

NO may play a major role in nervous system morphogenesis and plasticity and may be involved in activity-dependent establishment of connections in both developing and regenerating neurons.[63,66,67] Under developmental conditions, NO may trigger growth arrest, a process that at least in certain cell types, might involve inhibition of cdk2, a key regulator of the G1 and S phases of the cell cycle (see below). These antiproliferative effects of NO involve the repression of cyclin A reexpression as well as an induction of the cyclin-dependent kinase inhibitor p21^{Cip1}.[68] The high degree of coexpression of nNOS with p16^{INK4a},[69] indicates that further regulators of the G1-S-transition might be involved in the NO-induced cell-cycle-arrest or that additional mechanisms of proliferation and differentiation regulating mechanisms are activated in parallel in the course of neurodegeneration in AD. Thus NO serves as an inducer of cell-cycle-arrest, initiating the switch to cytostasis during differentiation,[63,65] a process that can alternatively lead to apoptosis.[68]

Although the molecular mechanism for the control of NO in proliferation, differentiation, cellular survival and death is not understood in detail, recent evidence indicates that activation of p21ras is critically involved in downstream signaling as a potential endogenous NO-redox sensitive effector molecule.[70,71] Endogenous NO and intermediates generated through oxidative stress can drive the Ras/MAPK cascade directly by direct activation of ras-GTPase activity.[66,70] NO might, thus, be a key mediator linking cellular activity to gene expression and long-lasting neuronal responses through activating p21ras by redox sensitive modulation[63] a process with potential implications both in development and degeneration.

In AD, nNOS is aberrantly expressed in potentially vulnerable neurons of the isocortex and entorhinal cortex. Since these neurons express nNOS prior to their affection by neurofibrillary degeneration,[69,72] transcriptional induction of nNOS might be an early event in the process of neurodegeneration. As expression of nNOS in AD is highly colocalized with p21ras,[69] an autocrine loop may exist within cells, whereby NO activates p21ras that in turn leads to cellular activation and stimulation of NOS expression.[70] The coexpression of NOS and p21ras in neurons vulnerable to neurofibrillary degeneration early in the course of AD might, thus, provide the basis for a feed back mechanism that might exacerbate the progression of neurodegeneration in a self propagating manner (Fig. 6). This self perpetuation of a process

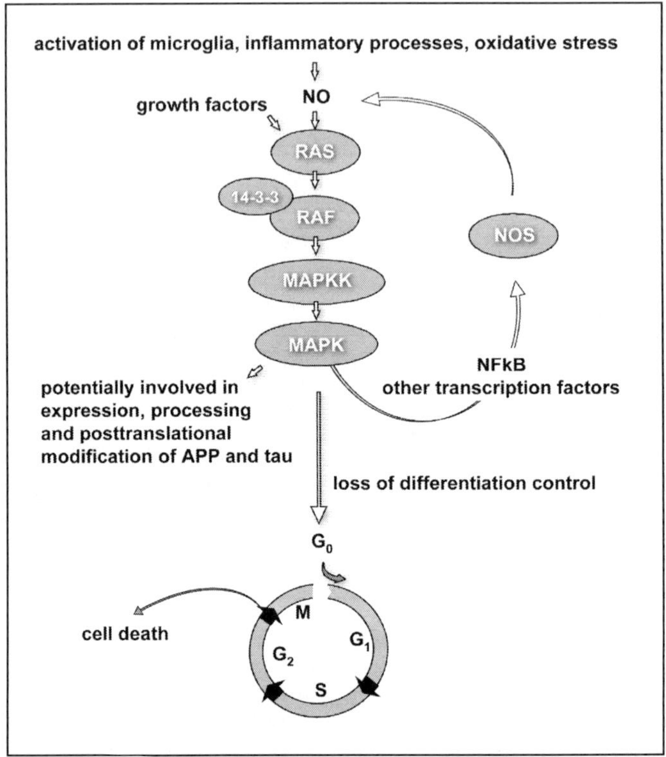

Figure 6. Synopsis of intracellular signaling events triggered by morphodysregulation in AD that involve an aberrant activation of p21ras/MAP-kinase signaling, a loss of differentiation control, the subsequent reentry and partial completion of the cell cycle and eventually result in cell death.

likely to be associated with limited prospects of physiological control and termination might be the critical switch converting two potentially neuroprotective mechanisms such as NO[73] and p21ras[74] dependent signaling into a disease process leading to slowly but continuously progressing neuronal death.

Brain-Morpho-Dysregulation and Replay of Developmental Mechanisms

Several of the genes and proteins and their posttranslational modification associated with AD have recently been found to play central roles in neuronal development, particular neuronal migration and axon extension.[75-79] Neurodegeneration, furthermore, is associated by a complex structural reorganisation that involves a replay of developmental programmes. The aberrant neuritic growth in AD, as a likely indication of defect synapse turnover, is accompanied by microtubular reorganization[42] associated with the reexpression of a number of developmentally regulated proteins involved in morphoregulation indicating that molecules, overexpressed in AD, might play a role in structural remodeling the adult brain.[77]

Based on this evidence, we have suggested some years ago that it is the process of continous synaptic reorganisation itself that becomes defective in AD.[2] In this pathogenetic process, a subset of neurons retaining a high degree of plasticity and which are presumably in a "labile state of differentiation", are forced into a condition of dedifferentiation that is characterized by an expression of developmental regulated genes, posttranslational modifications, and an accumulation of gene products to an extent that goes beyond those observed during regeneration. This replay of developmentally regulated mechanisms might be the endstage of disturbed structural brain self-organization and a slowly progressing "morpho-dysregulation". This process of dedifferentiation involves molecular events that, in dividing cell populations, would lead to cellular transformation and is, thus, not compatible with the state of a neuron being irreversibly blocked from the reentry into the cell cycle. It might, therefore, lead to neuronal death. From this hypothesis it can be predicted that those molecular events that are involved in neoplastic transformation might also play a key role in the pathomechanism of AD.[1] These mechanisms are notably a dysfunction of mitogenic signal transduction and cell-cycle control.

Cell-Cycle as an Alternative Effector Pathway of Plasticity-Related Signaling Mechanisms

Induction of the cell cycle by mitogenic compounds are mediated through cdk activation by cyclin D, the positive regulator of cdk4/6, and cyclin E, the positive regulator of cdk2, together with the repression of cdk-inhibitors of the cip/kip family (p21[Cip1], p27[Kip1] and p57[Kip2]) which are negative regulators of G1-phase cdks.[80] Cyclins D and E, moreover, are major downstream targets of signaling pathways involved in mediating structural plasticity and morphoregulation such as integrin-signaling, cadherin-signaling, wnt-signaling, LDL signaling and others. As a result, these pathways alternatively regulate plasticity and cell cycle activation.

Proteins that normally function to control cell cycle progression in actively dividing cells may play roles in the death of terminally differentiated postmitotic neurons.[81,82] Thus, a dysregulation of cdks and their regulating partners such as cyclins and cdk-inhibitors indicating an activation of the cell cycle in postmitotic neurons has been observed in AD,[8,10,75,83-92] cerebral ischemia and during trophic factor deprivation.[93-99] Correspondingly, pharmacological inhibitors of cell cycle or ectopic expression of cdkIs can protect neurons against death.[81,82,94,95,100 103]

Cyclin Dependent Kinases (cdks)

Cyclin dependent kinases (cdks) are Ser/Thr kinases that are only active in association with a regulatory partner (i.e., cyclin or other protein).[104,105] Cdks are critically involved in regulation of the cell cycle (cdk1, cdk2, cdk3, cdk4, cdk6, cdk7), as well as other functions such as transcription (cdk2, cdk7, cdk8, cdk9, cdk11), neurite outgrowth, neuron migration and neurotransmitter signaling (cdk4, cdk5, cdk11), differentiation (cdk2, cdk5, cdk6, cdk9) and cell death (cdk1, cdk2, cdk4, cdk5, cdk6, cdk11).

The orderly progression through the G1, S, G2 and M phases of the cell division cycle is driven by the sequential activation of cdks which is controlled through positive and negative regulators. The initial segment of the cell-cycle, the first gap (G1)-phase is the site of integration of mitotic signals that result from ligand dependent activation of both growth factor receptors and integrins[106] that both converge on the activation G1-cyclin-dependent kinases (cdks) cdk4/6 and cdk2.[107] Resulting cdk activities determine whether mitogenic signals are propagated downstream inducing phosphorylation of key substrates required for progression through G1 and entry into S-phase. The major function of G1-phase kinases is to phosphorylate the retinoblastoma protein (pRb).

Activity of these kinases is controlled at multiple levels, (i) from the accumulation of cyclins [Cyclin D1 linked to cdk4 or 6, and cyclin E linked to cdk2], (ii) at the level of assembly into a cyclin-cdk-complex, (iii) by phosphorylation and (iv) by their association with inhibitory proteins, the cyclin-dependent kinase inhibitors that can either block activation or block substrate/ATP access.[105,107]

The first wave of cyclin D-dependent kinase activity is followed in late G1 by an increase in cyclin E-cdk2 activity.[108,109] The cdk2-cyclin E complex is responsible for the G1/S transition but also regulates centrosome duplication. Unlike the D-type cyclin dependent kinases, assembly of cyclin E and cdk2 into active kinase is not mitogen-dependent. cdk2 reinforces cdk4 to complete Rb phosphorylation and induces the degradation of p27^{Kip1}. The cell cycle is now irreversibly committed to enter the S-phase. As cells transit the S-phase, proteolysis of cyclin E occurs. During S-phase, cdk2-cyclin A phosphorylates various substrates allowing DNA replication. At the S/G2 transition, cdk1 associates with cyclin A. In late G2, cdk1-cyclin B appears and triggers the G2/M transition, and cyclin A is degraded,[110] which resets the system and reestablishes the requirement for mitogenic cues to induce D-type cyclins for the next cycle.

The successive waves of cdk activation and inactivation are regulated through posttranslational modifications and subcellular translocations of positive and negative regulators, and are coordinated dependent on the completion of previous steps, through so-called checkpoint controls.[111] These checkpoints allow alternative decisions between further progression through the cycle, growth arrest or induction of apoptosis. Apoptosis as an alternative to cell cycle progression, is, thus, aimed at preventing cell proliferation. In proliferating cells, this might be a protective mechanism against transmission of nonreparable DNA damage. In highly differentiated cells, such as neurons, the mechanism might be preserved to prevent inappropriate generation of new cells that cannot easily be intergrated into functional circuitry.[112] An apoptotic response might, thus, be inherent to an inappropriate activation of the cell cycle.[113]

Several cyclin-dependent kinases critical for the progression through the cell cycle such as cdk1 (cdc2), cdk4 and cdk5 are deregulated in AD.[83,89,91,114-116] The cdk1 (cdc2) kinase, moreover, is able to phosphorylate tau protein at sites known to be phosphorylated in AD, and thus potentially contributes to the generation of PHF-tau.[115,117] APP, furthermore, is phosphorylated both in vitro and in intact cells by a cdk1 (cdc2)-like kinase in a cell cycle dependent manner which is associated with altered production of potentially amyloidogenic fragments containing the entire β/A4-domain.[118] Aβ in turn might potentially act as proliferating signal, driving cultured rat primary neurons into the cell cycle,[119] although in vivo, this has not been replicated.[120]

Cyclins

When quiescent cells enter the G1-phase, genes encoding D-type cyclins (D1, D2, D3) are induced in response to mitogenic signals. During progression through G1, the level of D-type cyclins increases, and, in a mitogen-regulated manner, these proteins associate with, and activate their catalytic partners, cdk4 or cdk6.[121] Assembled cyclin D-cdk complexes enter the nucleus where they become phosphorylated by a cdk-activating kinase (CAK). The major function of D-type cyclins is to provide a link between mitogenic stimulation and the autonomous progressing cell cycle machinery. D-type cyclins are, thus, usually absent from the cell cycle that progress independently of the presence of mitogenic signals. Conversely, constitutive activation of cyclin D can contribute to oncogenic cell transformation.

The level of cyclin D1, a critical regulator of the transition from the G_0 to the G_1 phase of the cell cycle that acts through activation of cdk4 is increased in neurons prone to neurodegeneration in AD.[83,114] Cyclins other than D1 such as cyclins E and A involved in regulation of G_1/S-transition as well as cyclin B regulating G_2/M-transition[10,83,91,92,122] are also elevated.

Cdk Inhibitors

Activity of G1-phase cdks are negatively regulated by cdk-inhibitors which bind directly to cdks or to cdk-cyclin complexes.[80] According to structural features, Cdk inhibitors are grouped into two families, INK4 and Cip/Kip. The INK4 (inhibitors of cdk4) family currently consists of four members, $p16^{INK4a}$, $p15^{INK4b}$, $p18^{INK4c}$ and $p19^{INK4d}$, that commonly share the presence of ankyrin repeats.[123] The INK4a locus encodes two independent but overlapping genes, $p16^{INK4a}$ and $p19^{ARF}$, the mouse homologue of $p14^{ARF}$ that act in partly overlapping pathways.[124] All these proteins can inhibit cyclin D associated kinases. INK4 proteins compete with D-type cyclins for binding to the cdk subunit. The INK4-family of cyclin-dependent kinase inhibitors might be involved in the regulation of pathways that control cell growth and proliferation as well as cell death. Deregulation of these cdki-proteins results in either uncontrolled proliferation and neoplastic transformation or activation of apoptosis.[102,125] The inhibitory action of the INK4 proteins is largely dependent on the presence of pRb in the cell.

Three proteins currently comprise the Cip/Kip family (cdk interacting protein / kinase inhibitory protein): $p21^{Cip1}$, $p27^{Kip1}$ and $p57^{Kip2}$. These proteins share a homologous inhibitory domain, which is necessary for binding of cdk4 and cdk2 complexes. Compared to the INK4-family, they have a wider inhibiting specificity, affecting the activities of cyclin-D-, E-, and A-dependent kinases. In vivo they preferentially act on cdk2 complexes. Through binding to cyclin D-cdk4 complexes, $p27^{Kip1}$ and $p21^{Cip1}$ can be sequestered without inhibiting the activity of the complex.[80] Kip/Cip-proteins might even be able to promote activation of cyclin D-cdk complexes by stabilizing the complexes and directing their nuclear translocation.[106] When mitogens are withdrawn and Cyclin D is downregulated, this pool of Kip/Cip-proteins is released, thereby inducing G1 phase arrest through inhibition of cyclin E-cdk2. $p21^{Cip1}$ also associates with PCNA, a subunit of DNA-polymerase delta which might be an additional mechanism through which $p21^{Cip1}$ can inhibit DNA synthesis.

$p16^{INK4a}$, a prominent representative of the INK4-family, involved in pathways for control of cell growth and proliferation[127] is also increased in AD as are other members of the INK4-family of the cyclin-dependent kinase inhibitors interacting with cdk4/6 such as $p15^{INK4b}$, $p18^{INK4c}$ and $p19^{INK4d}$, while alterations of $p21^{Cip1}$ and $p27^{Kip1}$ are less constant.[8,84,85]

Retinoblastoma Protein

G1-cdk activities determine whether mitogenic signals are propagated downstream resulting in phosphorylation of key substrates required for progression through G1. One of the key substrates is the retinoblastoma protein (Rb), an inhibitor of progression through G1. Activation of cyclin D-dependent-cdks initiate Rb phosphorylation in mid-G1 phase when cyclin E-cdk2 becomes active and completes this process by phosphorylating Rb on additional sites.[126,128] Cyclin A and B-dependent cdks activated later during the cell cycle maintain Rb in a hyperphosphorylated form. Resetting the pRb-E2F complex occurs in M-phase by dephosphorylation, probably by PP1.[129]

In quiescent cells, Rb is hypophosphorylated, and transcription factors such as E2F bind specifically to the hypophosphorylated form of pRb which act as a proliferation inhibitor.[130,131] Two other proteins, p130 and p107, are Rb-related members of the pRb "pocket protein" gene family. They are also substrates of the G1-phase cdks, share structural and biochemical properties and, like pRb, bind and modulate the activity of E2F transcription factors.[132]

Rb-E2F complexes bind to E2F-binding sites in E2F-responsive genes and repress transcription. Cdk-dependent phosphorylation of Rb (pRb) disrupts its association with E2F family members relieves pRb-mediated repression and allows for E2F-dependent transcription of several genes involved in cell-cycle and growth-regulation, including enzymes of DNA metabolism, protooncogenes, and cell-cycle regulatory proteins such as cyclins E and A that both are required to catalyze the G1/S transition.[126,133-135] Hypophosphorylated pRb also binds to other proteins, such as members of the abl family which seem to be closely implicated in adhesion-dependent growth control. The E2F-dependent induction of cyclin E, that in turn stimulates Rb phosphorylation through cdk2, provides a feedback loop that contributes to the G1/S transition. This event of pRb phosphorylation and release of E2F corresponds to progression through the restriction point R. This restriction point, R,[136] marks the point where cell cycle progression becomes independent of extracellular signals and cells become irreversibly committed to continued cell cycle progression.

In addition to regulation of E2F activity through phosphorylation of pocket proteins, direct phosphorylation of E2F also influences its activity. Both cyclin A-cdk2 and cyclin A-cdc2 can phosphorylate E2F-1 on multiple sites which might contribute to its downregulation in late S/G2 phase. Besides its involvement in cell growth and the cell cycle regulation, E2F also plays a role in apoptosis. Similar to growth regulation, E2F-mediated apoptosis is regulated by pocket proteins.[137]

In AD, increased immunoreactivity for hyperphosphorylated pRb and for E2F has been described.[138,139]

The Cell Cycle Integrates Intercellular and Intracellular Signaling and Links Development, Oncogenesis and Neurodegeneration

Cell Cycle Regulation through External "Positional" Cues

Activation of the cell cycle through growth factors also requires signals by the extracellular matrix. Entry into G1-phase and further progression towards the S-phase is a major downstream event of synergistic signaling by mitogenic compounds and integrin-dependent cell adhesion. All of the important mitogenic signaling cascades downstream of the Ras and Rho family small GTPases and the PI3-kinase-PKB/Akt pathway that control critical molecular switches such as induction of cyclin D1 resulting in activation of cdk4/6, the suppression of p21 and p27 inducing cdk2 activity[140,141] and the subsequent phosphorylation of Rb are regulated by integrin mediated cell adhesion.[142] Some of these matrix related effects involve matrix-dependent organization of the cytoskeleton. Adhesion-dependent cell cycle progression involves both integrin binding to the ECM and integrin-mediated cytoskeletal organization.[143] An organized cytoskeleton is, thus a requirement throughout the mitogen-dependent portion of the G1-phase; and the adhesion-dependent expression of cyclin D and the phosphorylation of pRb are mediated by the cytoskeleton. Cell adhesion to ECM is mediated by binding to cell surface integrin receptors, which activate intracellular signaling cascades and mediate tension-dependent changes in cell shape and cytoskeltal structure. Although growth control has focused on integrin and growth factor signaling, cell shape might play an equally critical role in cell cycle progression that act by subjugating the molecular machinery that regulates the G1/S transition.[144]

Adhesion to substratum has two separate effects on cell which both could underlie the anchorage requirement for cell cycle progression, i.e the initial adhesion event and the subsequent clustering of occupied integrins. Integrins might act as mechanoreceptors, transmitting mechanical information from the extracellular matrix to the cytoskeleton.[145,146] Alternatively, an organized cytoskeleton might be required to force integrins to remain clustered at focal contacts, and integrin clustering seems to be a prerequisite for integrin signaling.

Signaling through Ras and integrin regulates cyclinD-cdk4 and cylinE-cdk2 at different levels. (i) Expression of cyclin D1, often the rate-limiting step in the activation of cylinD-cdk4/6, critically dependent on synergistic signaling through the Ras/MAPK cascade, integrin, cadherin and wnt dependent signaling.[147] (ii) Assembly of newly synthesized cyclin D1 with cdk4 similarly depends on the Ras/MAPK cascade.[148] (iii) Turnover of D-type cyclins, morover, is Ras-dependent and mediated by the PI3-kinase-Akt-pathway that regulates the phosphory-lation of cyclin D1 by GSK-3β.[149] Inhibition of this pathway leads to cyclin D phosphoryla-tion, enhancing its nuclear export and the ubiquitin-dependent proteasomal degradation. (iv) Cell adhesion is a prerequisite for efficient down regulation the steady state levels of p21[Cip1] and/or p27[Kip1].[150] p27[Kip1] was first identified as a cdk2-inhibitory factor detected in contact-inhibited cells[151] and many factors, including cell-cell contact, induce accumulation of p27[Kip1].[152] Under these conditions, p19[INK4α] and p27[Kip1] cooperate to maintain differentiated neurons in a quiescent state. Adhesion lowers the levels of p21[Cip1] and p27[Kip1], thus permitting activation of cyclin E-cdk2 while high levels of p27[Kip1] persist in cells that were prevented from spreading by restriction of the size of the substratum, despite normal MAPK activation.[144]

Key intermediates of signaling of integrin-dependent cell adhesion upon cell cycle regula-tion are the integrin-linked kinase, that through inhibition of GSK-3β can also regulate both the expression and degradation of cyclin D1 and FAK that controls cell-cycle progression through G1 via JNK and through Ras - ERK1/2.[106]

Downstream G1 events, including phosphorylation of pRb leading to release of E2F-transcription factors from their complex with Rb and cyclin A expression also require both soluble mitogens and cell interaction with the extracellular matrix.[140,153] Integrin-mediated cell anchorage, furthermore, has a vital role in the control of apoptosis, a mechanism also dependent on FAK and Ras-dependent signaling as well as involving the wnt-1/β-catenin path-way. Control of anchorage dependent apoptosis, also coined 'anoikis',[154] may vary substan-tially from one cell lineage to another and Ras may fulfil both positive and negative regulatory function depending on the cell type.[140,155-157]

Cell Cycle Regulation through Oncogenes

Naturally, a variety of proto-oncogenes are connected to cell cycle regulation. Several oncoproteins are mutationally active forms of enzymes that act upstream in the MAP-kinase pathway, including the v-ErbB, ErbB2/Neu, and v-Src protein tyrosine kinases, Ras and Raf, indicating a critical role of persistent activation of the MAP-kinase pathway in oncogenesis. Small GTPases of the Ras-family which act as molecular switches in intracellular signaling regulate cellular proliferation and mediate the mitogenic response of a variety of growth factors and oncogenes.[48] Proto-oncogenes involved in the G_0/G_1 transition, such as *myc* and *ras*, are able to increase transcription of many immediate early genes including cyclin D1 and also to directly coorperate with cyclin D1 in transforming cells.[158] Mutations conferring constitutive Ras activation are found in nearly 30% of all human tumors.[159] Oncogenic *ras* promotes un-controlled mitogenesis independently of the presence of growth factors but itself is unable to transform cells. Constitutively-activated *ras*, moreover, not only regulates cellular proliferation, but also renders cells susceptible to apoptosis.[160-164]

Depending on the activation of other signaling pathways, activation of Ras signaling path-ways can also arrest the cell cycle rather than activate proliferation.[161,165] As a result, *ras* sup-presses oncogenic transformation and induces a condition phenotypically identical to prema-ture cellular senescence.[161,166] It selectively inhibits genes involved in mitosis, DNA replication, segregation, and repair, and at the same time upregulates genes associated with cell death and potentially involved in neurodegenerative disorders and tumorgenesis such as APP, serum amy-loid (SAA), tissue transglutaminase (t-TGase). This process also involves the upregulation of the cyclin dependent kinase inhibitors p21[Waf1/Cip1], p19[ARF] and p16[INK4α][161,167-172] and, thus, resembles changes seen in early AD.[7,8,55,84,85] Activation of the p21ras cascade, thus, plays an essential role in oncogenesis but also in cellular senescence as well as AD.

As we have shown previously, a high capacity of structural neuronal plasticity in the adult brain might neurons predispose to tangle formation in AD.[39,76,173] This high potential of neuroplasticity associated with the necessity of synaptic turnover and reorganisation might require properties inherent to both growth cones and synaptic connections.[174] Highly plastic neurons might, thus, retain 'immature' features and might not be 'fully differentiated' or 'truly postmitotic', i.e., arrested in G_0, an assumption supported by recent findings on the expression of cyclin B and E in hippocampal neurons of healthy elderly.[10,122]

It is, therefore, suggested that the reexpression of developmentally regulated genes, the induction of posttranslational modifications and accumulation of gene products to an extent which goes beyond that observed during regeneration and the aborted attempt of 'differentiated' neurons to activate the cell cycle, which apparently is a critical event in the pathomechanism of AD[2,9,10,76,83,85,88-92,114,122,173,175-177] is due to a loss of differentiation control that normally is involved in the regulation of neuronal plasticity. A direct link between cell cycle reactivation and cell death is supported by observations on the neuroprotective action of overexpression of the cyclin-dependent kinase inhibitor $p16^{INK4a}$ which locks neurons in a differentiated stage and prevents cell cycle reentry.[35]

It might, thus, be a 'labile fixation' of plastic neurons in G_0 which allows for ongoing morphoregulatory processes after development is completed. The delicate balance, however, between G_0-arrest and G_1-entry might be prone to a variety of potential disturbances during the lifetime of an individual. Morphodysregulation in AD, accompanied by aberrancies in intracellular mitogenic signaling might, thus, be a slowly progressing dysfunction that eventually overides this differentiation control and results in dedifferentiation, a condition in conflict with the otherwise 'mature' background of the nervous system. Cell-cycle and differentiation control might thus provide the link between structural brain self-organization and neurodegeneration[1,2] both of which in the human brain have reached a phylogenetic level unique in nature.

Conclusions

Taken together, in multicellular organisms, cell number is regulated spatially by extracellular signals through cell interactions controlling proliferation and survival in local neighbourhoods. Instructions from neighbouring cells can induce cell proliferation, differentiation or death. These stimuli include cell-cell and cell-ECM adhesion, growth factors, cytokines, neuropeptides and mechanical factors. Signals from G-protein-coupled receptors, tyrosine kinase receptors and integrins cooperate to integrate information from multiple stimuli that regulate cell cycle progression. To allow for a regulation of these processes a tight link is necessary between cell attachement mechanisms and control of proliferation and differentiation, i.e., the cell cycle machinery.

Contrary to most other cells that make up a multicellular organism, neurons use the molecular circuitry developed to sense their relationship to other cells to reorganize their connectivity according to the requirements for information processing within a cellular network. This puts neurons on the permanent risk to erroneously convert signals derived from plastic synaptic changes into positional cues that will activate the cell cycle. Maintaining neurons in a differentiated but still highly plastic phenotype will, thus, be the challenge to prevent neurodegeneration.

References

1. Heintz N. Cell-death and the cell-cycle - a relationship between transformation and neurodegeneration. Trends Biochem Sci 1993; 18:157-159.
2. Arendt TH. Neuronal dedifferentiation and degeneration in Alzheimer's disease. Biol Chem Hoppe-Seyler 1993; 374:911-912.
3. Ramón Y, Cajal S. Degeneration and regeneration of the nervous system. London: Oxford University Press, 1928.
4. Bouman L. Senile plaques. Brain 1934; 57:128-142.
5. Scheibel ME, Scheibel AB. Structural changes in the aging brain. Aging Vol. 1 Clinical, Morphologic and Neurochemical Aspects in the Aging Central Nervous System, 1975; 11-37.

6. Woolf NJ, Butcher LL. Dysdifferentiation of structurally plastic neurons initiates the pathologic cascade of Alzheimer's disease: Toward a unifying hypothesis. In: Steriade M, Biesold D, eds. Brain Cholinergic Systems. Oxford University Press, 1990:387-438.

7. Gärtner U, Holzer M, Heumann R et al. Induction of p21[ras] in Alzheimer pathology. Neuroreport 1995; 6:1441-1444.

8. Arendt TH, Rödel L, Gärtner U et al. Expression of the cyclin-dependent kinase inhibitor p16 in Alzheimer's disease. Neuroreport 1996; 7:3047-3049.

9. Vincent I, Rosado M, Davies P. Mitogenic mechanisms in Alzheimer's disease? J Cell Biol 1996; 132:413-425.

10. Nagy Z, Esiri MM, Cato AM et al. Cell cycle markers in the hippocampus in Alzheimer's disease. Acta Neuropathologica 1997; 94:6-15.

11. Flechsig P. Gehirn und Seele. Druck von Alexander Edelmann, Leipzig 1896.

12. Vogt C, Vogt O. Importance of neuroanatomy in the field of neuropathology. Neuro 1951; 205-218.

13. Ashford JW, Jarvik L. Alzheimer's disease: Does neuron plasticity predispose to axonal neurofibrillary degeneration? N Engl J Med 1985; 313:388-389.

14. Butcher LL, Woolf NJ. Neurotrophic agents may exacerbate the pathologic cascade of Alzheimer's disease. Neurobiol Aging 1989; 10:557-570.

15. Cotman CW, Anderson KJ. Synaptic plasticity and functional stabilization in the hippocampal formation: Possible role in Alzheimer's disease. Adv Neurol 1988; 47:313-335.

16. DeWitt DA, Silver J. Regenerative failure: A potential mechanism for neuritic dystrophy in Alzheimer's disease. Exp Neurol 1996; 142:103-110.

17. DiPatre PL. Cytoskeletal alterations might account for the phylogenetic vulnerability of the human brain to Alzheimer's disease. Med Hypotheses 1991; 54:103-170.

18. Flood DG, Coleman PD. Hippocampal plasticity in normal aging and decreased plasticity in Alzheimer's disease. Prog Brain Res 1990; 83:435-443.

19. Foster TC. Involvement of hippocampal synaptic plasticity in age-related memory decline. Brain Res Rev 1999; 30:236-249.

20. Geddes JW, Cotman CW. Plasticity in Alzheimer's disease: Too much or not enough? Neurobiol Aging 1991; 12:330-333.

21. Kondo M, Imahori Y, Mori S et al. Aberrant plasticity in Alzheimer's disease. Neuroreport 1999; 10:1481-1484.

22. Masliah E, Mallory M, Hansen L et al. Patterns of aberrant sprouting in Alzheimer's disease. Neuron 1991; 6:729-739.

23. Mesulam MM. Neuroplasticity failure in Alzheimer's disease: Bridging the gap between plaques and tangles. Neuron 1999; 24:521-529.

24. Mirmiran M, van Someren EJ, Swaab DF. Is brain plasticity preserved during aging and in Alzheimer's disease? Behav Brain Res 1996; 78:43-48.

25. Neill D. Alzheimer's disease: Maladaptive synaptoplasticity hypothesis. Neurodegeneration 1995; 4:217-232.

26. Phelps CH. Neural plasticity in aging and Alzheimer's disease: Some selected comments. Prog Brain Res 1990; 86:3-9.

27. Roberts GW, Nash M, Ince PG et al. On the origin of Alzheimer's disease: A hypothesis. Neuroreport 1993; 4:7-9.

28. Swaab DF. Brain aging and Alzheimer's disease, "wear and tear" versus "use it or lose it". Neurobiol Aging 1991; 12:317-324.

29. Teter B, Ashford W. Neuroplasticity in Alzheimer's disease. J Neurosci Res 2002; 70:402-437.

30. Walsh TJ, Opello KD. Neuroplasticity, the aging brain, and Alzheimer's disease. Neurotoxicology 1992; 13:101-110.

31. Ben-Ari Y, Represa A. Brief seizure episodes induce long-term potentiation and mossy fibre sprouting in the hippocampus. Trends Neurosci 1990; 13:312-318.

32. Jeffery KJ, Reid IC. Modifiable neuronal connections: An overview for psychiatrists. Am J Psychiatry 1997; 154:156-164.

33. McEachern JC, Shaw CA. An alternative to the LTP orthodoxy: A plasticity-pathology continuum model. Brain Res Rev 1996; 22:51-92.

34. Abbott LF, Nelson SB. Synaptic plasticity: Taming the beast. Nat Neurosci 2000; 3(Suppl):1178-1183.
35. Arendt TH. Synaptic plasticity and cell cycle activation in neurons are alternative effector pathways: The Dr. Jekyll and Mr. Hyde Concept of Alzheimer's disease or The yin and yang of Neuroplasticity. Progress in Neurobiology 2003; 71:83-248.
36. Arendt TH, Zvegintseva HG, Leontovich TA. Dendritic changes in the basal nucleus of Meynert and in the diagonal band nucleus in Alzheimer's disease- A quantitative golgi investigation. Neuroscience 1986; 19:1265-1278.
37. Arendt TH, Brückner MK. Perisomatic sprouts immunoreactive for nerve growth factor receptor and neurofibrillary degeneration affect different neuronal populations in the basal nucleus in patients with Alzheimer's disease. Neurosci Lett 1992; 148:63-66.
38. Buell SJ, Coleman PD. Dendritic growth in the aged human brain and failure of growth in senile dementia. Science 1979; 206:854-856.
39. Arendt TH, Brückner MK, Bigl V et al. Dendritic reorganization in the basal forebrain under degenerative conditions and its defects in Alzheimer's disease. II. Ageing, Korsakoff's disease, Parkinson's disease, and Alzheimer's disease. J Comp Neurol 1995a; 351:189-222.
40. Arendt TH, Brückner MK, Bigl V et al. Dendritic reorganization in the basal forebrain under degenerative conditions and its defects in Alzheimer's disease. III. The basal forebrain compared to other subcortical areas. J Comp Neurol 1995b; 351:223-246.
41. Arendt TH, Marcova L, Bigl V et al. Dendritic reorganization in the basal forebrain under degenerative conditions and its defects in Alzheimer's disease. I. Dendritic organisation of the normal human basal forebrain. J Comp Neurol 1995d; 351:169-188.
42. McKee AC, Kowall NW, Kosik KS. Microtubular reorganization and dendritic growth response in Alzheimer's disease. Ann Neurol 1989; 26:652-659.
43. Fischer O. Miliare Necrosen mit drusigen Wucherungen der Neurofibrillen, eine regelmässige Veränderung der Hirnrinde bei seniler Demenz. Monatsschrift für Psychiatrie und Neurologie 1907; 22:361-372.
44. Seger R, Krebs EG. The MAPK signaling cascade. FASEB J 1995; 9:726-735.
45. Greenberg SM, Koo EH, Selkoe J et al. Secreted beta amyloid precursor protein stimulates mitogen activated protein kinase and enhances tau phosphorylation. Proc Natl Acad Sci USA 1994; 91:7104-7108.
46. Mills J, Charest DL, Lam F et al. Regulation of amyloid precursor protein catabolism involves the mitogen-activated protein kinase signal transduction pathway. J Neurosci 1997; 17:9415-9422.
47. Sadot E, Jaaro H, Seger R et al. Ras-signaling pathways: Positive and negative regulation of tau expression in PC12 cells. J Neurochem 1998a; 70:428-431.
48. Bourne HR, Sanders DA, McCormick F. The GTPase superfamily: A conserved switch for diverse cell functions. Nature 1990; 348:125–132.
49. Stokoe D, MacDonald SG, Cadwallader K et al. Activation of Raf as a result of recruitment to the plasma membrane. Science 1994; 264:1463-1467.
50. Arendt TH, Gärtner U, Seeger G et al. Neuronal activation of Ras regulates synaptic connectivity. Europ J Neurosci 2004; 19:2953-2966.
51. Borasio GD, John J, Wittinghofer A et al. Ras p21 protein promotes survival and fiber outgrowth of cultured embryonic neurons. Neuron 1989; 2:1087-1096.
52. Phillips LL, Belardo ET. Increase of c-fos and ras oncoproteins in the denervated neuropil of the rat dentate gyrus. Neuroscience 1994; 58:503-514.
53. Holzer M, Gärtner U, Klinz FJ et al. Activation of mitogen activated protein kinase cascade (MAPK)-cascade and phosphorylation of cytoskeletal proteins after neuron-specific activation of p21ras. I. MAPK-cascade. Neuroscience 2001; 105:1031-1040.
54. Holzer M, Rödel L, Seeger G et al. Activation of mitogen activated protein kinase (MAPK)-cascade and phosphorylation of cytoskeletal proteins after neuron-specific activation of p21ras. II. Cytoskeletal proteins and dendritic morphology. Neuroscience 2001b; 105:1040-1054.
55. Gärtner U, Holzer M, Arendt TH. Elevated expression of p21^{ras} is an early event in Alzheimer's disease and precedes neurofibrillary degeneration. Neuroscience 1999; 91:1-5.

56. Arendt TH, Holzer M, Großmann A et al. Increased expression and subcellular translocation of the mitogen-activated protein kinase kinase and mitogen-activated protein kinase in Alzheimer's disease. Neuroscience 1995c; 68:5-18.
57. Boulton TG, Nye SH, Robbins DJ et al. ERKs: A family of protein-serine/threonine kinases that are activated and tyrosine phosphorylated in response to insulin and NGF. Cell 1991; 65:663-675.
58. Bogoyevitch MA, Court NW. Counting on mitogen-activated protein kinases ERKs 3, 4, 5, 6, 7 and 8. Cellular signalling 2004; 16:1345-1354.
59. Martin KC, Michael D, Rose JC et al. MAP kinase translocates into the nucleus of the presynaptic cell and is required for long-term facilitation in Aplysia. Neuron 1997b; 18:899-912.
60. Impey S, Obrietan K, Storm DR. Making new connections: Role of ERK/MAP kinase signaling in neuronal plasticity. Neuron 1999; 23:11-14.
61. Drewes G, Lichtenberg-Kraag B, Doring F et al. Mitogen activated protein (MAP) kinase transforms tau protein into an Alzheimer-like state. EMBO J 1992; 11:2131-2138.
62. Goedert M, Cohen ES, Jakes R et al. p42 MAP kinase phosphorylation sites in microtubule-associated protein tau are dephosphorylated by protein phosphatase 2A1 - implication for Alzheimer's disease. FEBS Lett 1992; 312:95-99.
63. Dawson TM, Sasaki M, Gonzales-Zulueta M et al. Regulation of neuronal nitric oxide synthase and identification of novel nitric oxide signaling pathways. Progress in Brain Res 1998; 118:3-11.
64. Peunova N, Enikolopov G. Amplification of calcium-induced gene transcription by nitric oxide in neuronal cells. Nature 1993; 364:450-453.
65. Peunova N, Enikolopov G. Nitric oxide triggers a switch to growth arrest during differentiation of neuronal cells. Nature 1995; 375:68-73.
66. Bogdan C. Nitric oxide and the regulation of gene expression. Trends Cell Biol 2001; 11:66-75.
67. Lüth H-J, Arendt TH. Oxyde Nitrique et maladie d'Alzheimer. Alzheimer Actualités 1998a; 132:6-11.
68. Gansauge S, Nussler AK, Berger HG et al. Nitric oxide-induced apoptosis in human pancreatic carcinoma cell lines is associated with a G1-arrest and increase of the cyclin-dependent kinase inhibitor p21WAF1/CIP1. Cell Growth Differ 1998; 9:611-617.
69. Lüth H-J, Holzer M, Gertz H-J et al. Aberrant expression of nNOS in pyramidal neurons in Alzheimer's disease is highly colocalized with p21ras and p16INK4a. Brain Research 2000; 852:45-55.
70. Lander HM, Hajjar DP, Hempstead BL et al. A molecular redox switch on p21 (ras). Structural basis for the nitric oxide-p21 (ras) interaction. J Biol Chem 1997; 272:4323-4326.
71. Yun HY, Gonzalez-Zulueta M, Dawson VL et al. Nitric oxide mediates N-methyl-D-aspartate receptor-induced activation of p21 ras. Proc Natl Acad Sci USA 1998; 95:5773-5778.
72. Lüth H-J, Arendt TH. Nitric oxide and Alzheimer's disease. J Brain Research 1998b; 39:245-251.
73. Farinelli SE, Park DS, Green LA. Nitric oxide delays the death of trophic factor-deprived PC12 cells and sympathetic neurons by a cGMP-mediated mechanism. J Neurosci 1996; 16:2325-2334.
74. Heumann R, Goemans C, Bartsch D et al. Transgenic activation of ras in neurons promotes hypertrophy and protects from lesion-induced degeneration. Journal of Cell Biology 2000; 151:1537-1548.
75. Arendt TH. Alzheimer's disease as a loss of differentiation control in a subset of neurons that retain immature features in the adult brain. Neurobiology of Aging 2000; 21:783-796.
76. Arendt TH. Disturbance of neuronal plasticity is a critical pathogenetic event in Alzheimer's disease. Int J Dev Neurosci 2001a; 19:231-245.
77. Arendt TH. Alzheimer's disease as a disorder of mechanisms underlaying structural brain self-organization. Commentary Neuroscience 2001b; 102:723-765.
78. Bothwell M, Giniger E. Alzheimer's disease: Neurodevelopment converges with neurodegeneration. Cell 2000; 102:271-273.
79. Mehler MF, Gokhan S. Developmental mechanisms in the pathogenesis of neurodegenerative diseases. Progr Neurobiol 2001; 63:337-363.
80. Sherr CJ, Roberts JM. Inhibitors of mammalian G1 cyclin-dependent kinases. Genes Dev 1995; 9:1149-1163.
81. Farinelli SE, Greene LA. Cell cycle blockers mimosine, ciclopirox, and deferoxamine prevent the death of PC12 cells and postmitotic sympathetic neurons after removal of trophic support. J Neurosci 1996; 16:1150-1162.

82. Park DS, Farinelli SE, Greene LA. Inhibitors of cyclin-dependent kinases promote survival of post-mitotic neuronally differentiated PC12 cells and sympathetic neurons. J Biol Chem 1996; 271:8161-8169.

83. Arendt TH, Holzer M, Stöbe A et al. Activated mitogenic signalling induces a process of de-differentiation in Alzheimer's disease that eventually results in cell death. Ann NY Acad Sci 2000; 920:249-255.

84. Arendt TH, Holzer M, Gärtner U. Neuronal expression of cycline dependent kinase inhibitors of the INK4 family in Alzheimer's disease. J Neural Transm 1998b; 105:949-960.

85. Arendt TH, Holzer M, Gärtner U et al. Aberrancies in signal transduction and cell cycle related events in Alzheimer's disease. J Neural Transmission Suppl 1998c; 54:147-158.

86. Dranovsky A, Vincent I, Gregori L et al. Cdc2 phosphorylation of nucleolin demarcates mitotic stages and Alzheimer's disease pathology. Neurobiol Aging 2001; 22:517–528.

87. Jordan-Sciutto KL, Morgan K, Bowser R. Increased cyclin G1 immunoreactivity during Alzheimer's disease. J Alzheimers Dis 1999; 1:409–417.

88. McShea A, Harris PL, Webster KR et al. Abnormal expression of the cell cycle regulators P16 and CDK4 in Alzheimer's disease. Am J Pathol 1997; 150:1933-1939.

89. Nagy Z, Esiri MM, Smith AD. The cell division cycle and the pathophysiology of Alzheimer's disease. Neuroscience 1998; 87:731-739.

90. Smith TW, Lippa CF. Ki-67 immunoreactivity in Alzheimer's disease and other neurodegenerative disorders. J Neuropathol Exp Neurol 1995; 54:297-303.

91. Vincent I, Jicha G, Rosado M et al. Aberrant expression of mitogenic Cdc2/cyclin B1 kinase in degenerating neurons of Alzheimer's disease brain. J Neuroscience 1997; 17:3588-3598.

92. Vincent I, Jicha G, Rosado M et al. Aberrant expression of mitotic cdc2/cyclin B1 kinase in degenerating neurons of Alzheimer's disease brain. J Neurosci 1997; 588–598.

93. Gill JS, Windebank AJ. Cisplatin-induced apoptosis in rat dorsal root ganglion neurons is associated with attempted entry into the cell cycle. J Clin Invest 1998; 101:2842-2850.

94. Katchanov J, Harms C, Gertz K et al. Mild cerebral ischemia induces loss of cyclin-dependent kinase inhibitors and activation of cell cycle machinery before delayed neuronal cell death. J Neurosci 2001; 21:5045-5053.

95. Osuga H, Osuga S, Wang F et al. Cyclin-dependent kinases as a therapeutic target for stroke. Proc Natl Acad Sci USA 2000; 97:10254-10259.

96. Sakurai M, Hayashi T, Abe K et al. Cyclin D1 and Cdk4 protein induction in motor neurons after transient spinal cord ischemia in rabbits. Stroke 2000; 31:200-207.

97. Tomasevic G. Changes in proliferating cell nuclear antigen, a protein involved in DNA repair, in vulnerable hippocampal neurons following global cerebral ischemia. Brain Res Mol Brain Res 1998; 60:168–176.

98. van Lookeren Campagne M, Gill R. Increased expression of cyclin G1 and p21WAF/CIP1 in neurons following transient forebrain ischemia: Comparison with early DNA damage. J Neurosci Res 1998; 53:279-296.

99. Timsit S, Rivera S, Ouaghi P et al. Increased cyclin D1 in vulnerable neurons in the hippocampus after ischaemia and epilepsy: A modulator of in vivo programmed cell death? Eur J Neurosci 1999; 11:263-278.

100. Giovanni A, Wirtz-Brugger F, Keramaris E et al. Involvement of cell cycle elements, cyclin-dependent kinases, pRB, and EF2-DP, in β-amyloid induced neuronal death. J Biol Chem 1999; 274:19011-19016.

101. Knockaert M, Grenngard P, Meijer L. Pharmacological inhibitors of cyclin-dependent kinases. TRENDS in Pharmacological Sciences 2002; 23:417-418.

102. Kranenburg O, van der Eb A, Zantema A. Cyclin D1 is an essential mediator of apoptotic neuronal cell death. EMBO J 1996; 15:46-54.

103. Padmanabhan J. Role of cell cycle regulatory proteins in cerebellar granule neuron apoptosis. J Neurosci 1997; 19:8747-8756.

104. Morgan D. Cyclin-dependent kinases: Engines, clocks, and microprocessors. Annu Rev Cell Dev Biol 1997; 13:261-292.

105. Pavletich NP. Mechanisms of cyclin-dependent kinase regulation: Structures of Cdks, their cyclin activators, and Cip and INK4 inhibitors. J Mol Biol 1999; 287:821-828.
106. Assoian RK, Schwartz MA. Coordinate signaling by integrins and receptor tyrosine kinases in the regulation of G1 phase cell-cycle progression. Curr Opin Genet Dev 2001; 11:48-53.
107. Sherr CJ. G_1 phase progression: Cycling on cue. Cell 1994; 79:551-555.
108. Dulic V, Lees E, Reed SI. Association of human cyclin E with a periodic G1-S phase protein kinase. Science 1992; 257:1958-1961.
109. Koff A, Giordano A, Desai D et al. Formation and activation of a cyclin E-cdk2 complex during the G1 phase of the human cell cycle. Science 1992; 257:1689-1694.
110. Peters JM. SCF and APC: The Yin and Yang of cell cycle regulated proteolysis. Curr Opin Cell Biol 1998; 10:759-768.
111. Lew DJ. Cell-cycle checkpoints that ensure coordination between nuclear and cytoplasmic events in Saccharomyces cerevisiae. Curr Opin Gen Dev 2000; 10:47-53.
112. Copani A, Uberti D, Sortino MA et al. Activation of cellcycle-associated proteins in neuronal death: A mandatory of dispensable path? Trends Neurosci 2001; 24:25-31.
113. Evan G, Littlewood T. A matter of life and death. Science 1998; 281:1317-1322.
114. Busser J, Geldmacher DS, Herrup K. Ectopic cell cycle proteins predict the sites of neuronal cell death in Alzheimer's disease brain. J Neurosci 1998; 18:2801-2807.
115. Ledesma MD, Correas I, Avila J et al. Implication of brain cdc2 and MAP2 kinases in the phosphorylation of tau protein in Alzheimer's disease. FEBS Lett 1992; 308.218-224.
116. Patrick GN, Zukerberg L, Nikolic M et al. Conversion of p35 to p25 deregulates Cdk5 activity and promotes neurodegeneration. Nature 1999; 402:615-622.
117. Kobayashi S, Ishiguro K, Omori A et al. A cdc2-related kinase PJSALRE/cdk5 is homologous with the 30 kDa subunit of tau protein kinase II, a proline-directed protein kinase associated with microtubule. FEBS Lett 1993; 335:171-175.
118. Suzuki T, Oishi M, Marshak DR et al. Cell cycle-dependent regulation of the phosphorylation and metabolism of the Alzheimer amyloid precursor protein. EMBO J 1994; 13:1114-1122.
119. Copani A, Condorelli F, Caruso A et al. Mitotic signaling by beta-amyloid causes neuronal death. FASEB J 1999; 13:2225-2234.
120. Gärtner U, Brückner MK, Krug S et al. Amyloid deposition in APP 23 mice is associated with the expression of cyclins in astrocytes but not in neurons. Acta Neuropath 2003; 106:535-544.
121. Vidal A, Koff A. Cell-cycle inhibitors: Three families united by a common cause. Gene 2000; 247:1-15.
122. Smith MZ, Nagy Z, Esiri MM. Cell cycle-related protein expression in vascular dementia and Alzheimer's disease. Neurosci Lett 1999; 271:45-48.
123. Serrano M, Hannon GJ, Beach D. A new regulatory motif in cell-cycle control causing specific inhibition of cyclin D/CDK4. Nature 1993; 366:704-707.
124. Carnero A, Hudson JD, Price CM et al. p16INK4A and p19ARF act in overlapping pathways in cellular immortalization. Nat Cell Biol 2000; 2:148-155.
125. Freeman RS, Estus S, Johnson EM. Analysis of cell cycle related gene expression in postmitotic neurons. Selective induction of cyclin D1 during programmed cell death. Neuron 1994; 12:343-355.
126. Cheng M, Olivier P, Diehl JA et al. The p21(Cip1) and p27(Kip1) CDK ‚inhibitors‘ are essential activators of cyclin D-dependent kinases in murine fibroblasts. EMBO J 1999; 18:1571-1583.
127. Kamb A, Gruis NA, Weaver-Feldhaus J et al. A cell cycle regulator potentially involved in genesis of many tumor types. Science 1994; 264:436-440.
128. Lundberg AS, Weinberg RA. Functional inactivation of the retinoblastoma protein requires sequential modification by at least two distinct cyclin-cdk complexes. Mol Cell Biol 1998; 18:753-761.
129. Ludlow JW, Glendening CL, Livingston DM et al. Specific enzymatic dephosphorylation of the retinoblastoma protein. Mol Cell Biol 1993; 13:367-372.
130. Luo RX, Postigo AA, Dean DC. Rb interacts with histone deacetylase to repress transcription. Cell 1998; 92:463-473.
131. Weinberg RA. The retinoblastoma protein and cell cycle control. Cell 1995; 81:323-330.
132. Ewen ME. Regulation of the cell cycle by the Rb tumor suppressor family. Results Probl Cell Differ 1998; 22:149-179.

133. Black AR, Azizkhan-Clifford J. Regulation of E2F: A family of transcription factors involved in proliferation control. Gene 1999; 237:281-302.

134. Dyson N. The regulation of E2F by pRB-family proteins. Genes Dev 1998; 12:2245-2262.

135. Nevins JR. Toward an understanding of the functional complexity of the E2F and retinoblastoma families. Cell Growth Differ 1998; 9:585-593.

136. Pardee AB. G_1 events and regulation of cell proliferation. Science 1989; 246:603-608.

137. Shan B, Durfee T, Lee WH. Disruption of RB/E2F-1 interaction by single point mutations in E2F-1 enhances S-phase entry and apoptosis. Proc Natl Acad Sci USA 1996; 93:679-684.

138. Jordan-Sciutto KL, Malaiyandi LM, Bowser R. Altered distribution of cell cycle transcriptional regulators during alzheimer disease. Neuropathol Exp Neurol 2002; 61:358-367.

139. Hoozemans JJ, Veerhuis R, Rozemuller AJ et al. Neuronal COX-2 expression and phosphorylation of pRb precede p38 MAPK activation and neurofibrillary changes in AD temporal cortex. Neurobiol Dis 2004; 15:492-499.

140. Assoian RK. Anchorage-dependent cell cycle progression. J Cell Biol 1997; 136:1-4.

141. Wu RC, Schönthal AH. Activation of p53-p21waf1 pathway in response to disruption of cell-matrix interactions. J Biol Chem 1997; 272:29091-29098.

142. Danen EHJ, Yamada KM. Fibronectin, inegrins, and growth control. J of Cell Physiology 2001; 189:1-13.

143. Assoian RK, Zhu X. Cell anchorage and the cytoskeleton as partners in growth factor dependent cell cycle progression. Curr Opin Cell Biol 1997; 9:93-98.

144. Huang S, Chen CS, Ingber DE. Control of cyclin D1, p27(Kip1), and cell cycle progression in human capillary endothelial cells by cell shape and cytoskeletal tension. Mol Biol Cell 1998; 9:3179-3193.

145. Hansen LK, Mooney DJ, Vacanti JP et al. Integrin binding and cell spreading on extracellular matrix act at different points in the cell cycle to promote hepatocyte growth. Mol Biol Cell 1994; 5:967-975.

146. Wang N, Butler JP, Ingber DE. Mechanotransduction across the cell surface and through the cytoskeleton. Science 1993; 260:1124-1127.

147. Roovers K, Assoian RK. Integrating the MAP kinase signal into the G1 phase cell cycle machinery. Bioessays 2000; 22:818-826.

148. Cheng M, Sexl V, Sherr CJ et al. Assembly of cyclin D-dependent kinase and titration of p27Kip1 regulated by mitogen-activated protein kinase kinase (MEK1). Proc Natl Acad Sci USA 1998; 95:1091-1096.

149. Diehl JA, Cheng M, Roussel MF et al. Glycogen synthase kinase-3beta regulates cyclin D1 proteolysis and subcellular localization. Genes Dev 1998; 12:3499-3511.

150. Welsh CF, Roovers K, Villanueva J et al. Timing of cyclin D1 expression within G1 phase is controlled by Rho. Nature Cell Biology 2001; 3:950–957.

151. Polyak K, Lee MH, Erdjument-Bromage H et al. Cloning of p27Kip1, a cyclin-dependent kinase inhibitor and a potential mediator of extracellular antimitogenic signals. Cell 1994b; 78:59-66.

152. Hengst L, Reed SI. Inhibitors of the Cip/Kip family. Curr Top Microbiol Immunol 1998; 227:25-41.

153. Aplin AE, Howe AK, Juliano RL. Cell adhesion molecules, signal transduction and cell growth. Curr Opin Cell Biol 1999; 11:737-744.

154. Frisch SM, Francis H. Disruption of epithelial cell-matrix interactions induces apoptosis. J Cell Biol 1994; 124:619-626.

155. Frisch SM, Ruoslahti E. Integrins and anoikis. Curr Opin Cell Biol 1997; 9:701-706.

156. McGill G, Shimamura A, Bates RC et al. Loss of matrix adhesion triggers rapid transformation-selective apoptosis in fibroblasts. J Cell Biol 1997; 138:901-911.

157. Meredith Jr JE, Schwartz M. Integrins, adhesion and apoptosis. Trends Cell Biol 1997; 7:146-150.

158. Seth A, Gonzalez FA, Gupta S et al. Signal transduction within the nucleus by mitogen-activated protein kinase. J Biol Chem 1992; 267:24796-24804.

159. Bos J. Ras oncogenes in human cancer: A review. Cancer Res 1989; 49:4682–4689.

160. Khokhlatchev A, Rabizadeh S, Xavier R et al. Identification of a novel ras-regulated proapoptotic pathway. Current Biology 2002; 12:253–265.

161. Serrano M, Lin AW, McCurrach ME et al. Oncogenic ras provokes premature cell senescence associated with accumulation of p53 and p16INK4a. Cell 1997; 88:593-602.

162. Frame S, Balmain A. Integration of positive and negative growth signals during ras pathway activation in vivo. Curr Opin Genet Dev 2000; 10:106–113.

163. Joneson T, Bar-Sagi D. Suppression of Ras induced apoptosis by the Rac GTPase. Mol Cell Biol 1999; 19:5892–5901.

164. Liou JS, Chen CY, Chen JS et al. Oncogenic Ras mediates apoptosis in response to protein kinase C inhibition through the generation of reactive oxygen species. J Biol Chem 2000; 275:39001-390011.

165. Ridley AJ, Paterson HF, Noble M et al. Ras-mediated cell cycle arrest is altered by nuclear oncogenes to induce Schwann cell transformation. EMBO J 1988; 7:1635-1645.

166. Lin AW, Barradas M, Stone JC et al. Premature senescence involving p53 and p16 is activated in response to constitutive MEK/MAPK mitogenic signaling. Genes Dev 1998; 12:3008-3019.

167. Ferbeyre G, de Stanchina E, Querido E et al. PML is induced by oncogenic ras and promotes premature senescence. Genes Dev 2000; 14:2015-2027.

168. Gottifredi V, Prives C. P53 and PML: New partners in tumor suppression. Trends Cell Biol 2001; 11:184-187.

169. Guo A, Salomoni P, Luo J et al. The function of PML in p53-dependent apoptosis. Nat Cell Biol 2000; 2:730-736.

170. Malumbres M, Perez De Castro I, Hernandez MI et al. Cellular response to oncogenic ras involves induction of the Cdk4 and Cdk6 inhibitor p15(INK4b). Mol Cell Biol 2000; 20:2915-2925.

171. Pearson M, Carbone R, Sebastiani C et al. PML regulates p53 acetylation and premature senescence induced by oncogenic Ras. Nature 2000; 406:207-210.

172. Zhong S, Salomoni P, Pandolfi PP. The transcriptional role of PML and the nuclear body. Nat Cell Biol 2000b; 2:E85-90.

173. Arendt TH, Brückner MK, Gertz HJ et al. Cortical distribution of neurofibrillary tangles in Alzheimer's disease matches the pattern of neurones that retain their capacity of plastic remodelling in the adult brain. Neuroscience 1998a; 83:991-1002.

174. Pfenninger KH, de la Houssaye BA, Helmke SM et al. Growth-regulated proteins and neuronal plasticity. A commentary Mol Neurobiol 1991; 5:143-151.

175. Masliah E, Mallory M, Alford M et al. Immunoreactivity of the nuclear antigen p105 is associated with plaques and tangles in Alzheimer's disease. Lab Invest 1993c; 69:562-569.

176. Nagy Z, Esiri MM, Smith AD. Expression of cell division markers in the hippocampus in Alzheimer's disease and other neurodegenerative conditions. Acta Neuropathol 1997b; 93:294-300.

177. Vincent I, Zheng JH, Dickson DW et al. Mitotic phosphoepitopes precede paired helical filaments in Alzheimer's disease. Neurobiol Aging 1998; 19:287-296.

178. Arendt TH. Dysregulation of neuronal differentiation and cell cycle control in Alzheimer's disease. J Neural Transm 2002; 62:77–85.

179. Fang F, Orend G, Watanabe N et al. Dependence of cyclin E-cdk2 kinase activity on cell anchorage. Science 1996; 271:499-502.

180. Hindley S, Juurlink BH, Gysbers JW et al. Nitric oxide donors enhance neurotrophin-induced neurite outgrowth through a cGMP-dependent mechanism. J Neurosci 1997; 15:427-439.

181. Motonaga K, Itoh M, Hirayama A et al. Up-regulation of E2F-1 in Down's syndrome brain exhibiting neuropathological features of Alzheimer-type dementia. Brain Res 2001; 905:250-253.

182. Resnitzky D. Ectopic expression of cyclin D1 but not cyclin E induces anchorage-independent cell cycle progression. Mol Cell Biol 1997; 17:5640-5647.

183. Sherr CJ, Roberts JM. CDK inhibitors: Positive and negative regulators of G1-phase progression. Genes Dev 1999; 13:1501-1512.

184. Sweetser DA, Kapur RP, Froelick GJ et al. Oncogenesis and altered differentiation induced by activated Ras in neuroblasts of transgenic mice. Oncogene 1997; 15:2783-2794.

185. Zindy F, Cunningham JJ, Sherr CJ et al. Postnatal neuronal proliferation in mice lacking Ink4d and Kip1 inhibitors of cyclin-dependent kinases. Proc Natl Acad Sci USA 1999; 96:13462-13467.

CHAPTER 2

Paved with Good Intentions:

The Link between Cell Cycle and Cell Death in the Mammalian Central Nervous System

Yan Yang and Karl Herrup

Cell Division

Cell division is among the most basic of biological processes. All life forms, from blue-green algae to human hepatocytes, ultimately depend for their survival on the ability of one cell to create two. In keeping with the centrality of this process, the component enzyme systems have been well conserved in evolution. But while unicellular creatures such as bacteria, protozoa and yeast are free to divide whenever the nutrient source is adequate, multicellular organisms must tightly regulate the division of their constituent cells if they are to maintain their correct size and shape. Given this requirement, it is not surprising that the activities of the various cell cycle enzymes are regulated by a large and complex network of gene products. Indeed since life itself depends on the existence of a vigorous cell division process, it is nearly axiomatic that complex organisms must have an equally robust series of mechanisms to hold the cell cycle in check. Nowhere is this need for cellular restraint more crucial than in the adult central nervous system.

The mitotic cell cycle has four recognized phases. G1 phase is a period of variable length during which a single cell grows to a point where cell division is called for. After the commitment to begin division is made, G1 is followed by S phase in which the DNA synthetic machinery replicates the cell's genetic material and chromosome number doubles. The cell then prepares for division in a period known as G2 phase. Finally, during M phase, the chromosomes move to opposite poles of the cell; the cytoplasm is split and two daughter cells emerge. The proteins that regulate this process are diverse in both form and function. The central players are a series of protein kinases known as cyclin dependent kinases, or Cdks. The activities of these proteins vary with the phase of the cycle according to their own state of phosphorylation as well as their binding to a series of regulatory proteins known as cyclins. Additional levels of regulation are added by a number of cell cycle inhibitors (such as the cyclin D inhibitors p16 and p27) that act by binding to and blocking the action of various cyclins. Further, peptide destruction (targeted by ubiquitination and effected by specific peptidases such as cdc25) is yet another level of both positive and negative cell cycle regulation.

Cell Cycle Mechanisms and Neuronal Cell Death, edited by Agata Copani and Ferdinando Nicoletti. ©2005 Eurekah.com and Kluwer Academic / Plenum Publishers.

Cell Cycle Events in the Cell Death Process

Until recently, cell division and cell death seemed not only separate, but polar opposite concepts. The former would seem a generative process that favors growth and development while the latter is a destructive event that favors atrophy and loss. A substantial body of evidence now suggests, however, that the two processes are intimately related and use many of the same mechanisms for their execution. The idea that reactivation of the machinery of cell division in a mature neuron might to death rather than division, though paradoxical, fits well with an oft-observed but poorly understood phenomenon: there are almost no examples of tumors that are founded by CNS neurons. With very few exceptions,* the cancers that we refer to as "brain" tumors originate from nonneuronal cells such as astrocytes, oligodendrocytes, and cells of the meninges.

Some of the first direct hints of the cell cycle/cell death linkage came from the analyses of transgenic mice in which oncogenes such as SV40 T-antigen were driven with neuron-specific promoters. When T-antigen is expressed under the control of a rhodopsin promoter, the result is the loss of photoreceptor cells.[1] When it is expressed under a N-methyltransferase promoter retinal amacrine and horizontal cells die.[2] Finally, when the Purkinje cell-specific promoter of the *pcp2* gene is used to drive T-antigen expression, cerebellar Purkinje cell death rather than cell division is the unexpected result. Careful study of this latter situation revealed that the dying Purkinje cells were labeled by the DNA precursor, bromodeoxyuridine (BrdU) before their demise.[3] This suggests that the T-antigen oncogene had successfully initiated a cell cycle but for some reason the Purkinje cell could not complete it. T-antigen functions in part by binding the endogenous cellular protein, retinoblastoma (RB).[4] RB is a nuclear protein that regulates a key point in the cascade of events that initiates cell division.[5,6] The activity of RB is normally inhibitory (hence its classification as a tumor suppressor gene) but it can be modulated through protein phosphorylation; higher levels of phosphorylation inactivate RB and release the cell cycle. By binding and sequestering RB, the effect of T-antigen is to functionally mimic the effects of phosphorylation.

In the same year that the T-antigen transgenics were published, several cell cycle labs announced the creation of engineered null mutations in the mouse *retinoblastoma* gene.[7-9] The three labs were undoubtedly expecting to find unregulated cell division in the embryo; instead all three reported the occurrence of massive amounts of cell death in the central nervous system. All three groups noted the implication of this discovery: loss of cell cycle control in a newly generated neuron leads to cell death. One might have imagined that the curious cooccurrence of cell cycle anomalies and neuronal cell death was limited to artificial genetically engineered systems and somehow perhaps to other functions of the RB protein itself. Yet, shortly after these findings were announced, Freeman et al[10] showed that the death of sympathetic neurons following trophic factor deprivation led to the up-regulation of cyclin D1 (a G1 phase cyclin). This finding was of particular interest as the message levels for most of the other genes examined decreased. Following these observations in PNS neurons, our laboratory investigated several instances of target-related cell death in the CNS.[11]

Taken together, these early findings suggest a model of cell division in the mature nervous system that is the central hypothesis of this chapter:

Once a neuroblast makes the commitment to stop dividing and begin differentiation, any event that forces it back into the cell cycle will result in its death. This prohibition against cell division begins early in development and persists for the life of the organism.

---

* Retinoblastomas, meduloblastomas and multiple endocrine neoplasias

The Cell Biology of Cell Cycle Induced Cell Death

Since these first studies there has been a growing recognition of the wide spread applicability of this principle. Tissue culture model systems have been developed and have provided some of the most detailed evidence establishing a linkage between cell cycle and cell death. These data come from analyses of cell lines such as PC12 cells[12-16] as well as primary neuronal cultures.[13,15,17-20] The role of cell cycle processes is noted in a number of different experimental situations in which neuronal cell death is observed. For example, trophic factor withdrawal can induce a cycle-associated death in primary neurons and PC12 cells cultured in vitro.[15,17-19] Using pharmacological approaches, the laboratories of Greene and colleagues have shown that drugs block cell cycle advancement are efficient at preventing the death of both PC12 cells and sympathetic neurons.[17-20] Molecular genetic approaches have also been used. Dominant negative forms of the Cdk4 and Cdk6 proteins have been engineered and these too are effective in blocking the cell death process.[19] In addition, neurons subjected to DNA damaging agents such as UV irradiation or camptothecin (a topoisomerase inhibitor) require cyclin D1 and CDK4/6 activity to induce neuronal death.[21] These in vitro models are significant since many of them provide direct experimental evidence that, rather than merely being associated with cell death, an ectopic cell division is both necessary and sufficient to produce the death of neurons. An excellent review of this entire topic can be found in Liu and Green.[22]

The tissue culture studies and the observations in the RB mutants and T-antigen transgenics are strong evidence in support of the cell cycle and cell death connections. And this linkage is found a number of other in vivo situations of neuronal cell death. Recently, Chen et al[22a] have shown that in mice lacking the cell cycle inhibitor, p19[Ink4d], hair cells die post-natally. Their death is clearly cell cycle related as BrdU is incorporated into the normally post-mitotic cells. An additional example mentioned briefly above is the phenomenon of naturally occurring cell death. In normal development, most neurons go through a 'critical period' during which they have an acute dependency on contact with their target. If the contact is insufficient, the presynaptic neurons will rapidly die. It is believed that this pruning mechanism allows the different interconnected parts of the CNS to achieve a functional numerical balance that is optimum for adult function. One of the regions where target-related cell death has been studied most extensively is in the developing cerebellum. This is because many of the 'classic' neurological mutations of the mouse such as *lurcherr* and *staggere* lose large numbers of granule cell and inferior olive neurons due to the mutations' destructive effects on their target, the cerebellar Purkinje cell.[23,24] The presynaptic granule and olivary neurons are not themselves intrinsically compromised by either mutation; the Purkinje cells in both mutants are.[25-28] In both mutants the discovery of BrdU incorporation as well as immunocytochemical evidence for the reexpression of cell cycle proteins (Fig. 1A)[11] demonstrates that both neuronal cell types reenter an unscheduled cell cycle before their demise. This point is further underscored by the situation in the wild type mouse where even the normal pruning of the granule cell population that occurs during postnatal cerebellar development is found to proceed by the same reengagement of the cell cycle (Fig. 1B).[11]

The principle of cell cycling as a correlate of the cell death process has recently been extended further. When a neuron is deprived of oxygen, even for relatively short periods of time, it will die. This is true both in tissue culture and in vivo during disease related or experimentally induced ischemic incidents. Both situations have been examined for the evidence of cell cycle processes and in both situations such evidence has been found. In rats, focal ischemic insults as short as 30 min induce Cdk2 and Cdk4 as well as their associated cyclins, cyclin D1 and cyclin A and in response the levels of phosphorylated RB increase and E2F appears to be released.[29,30] As if to protect itself from the ill effects of the cell cycle, the Cdk inhibitors p21 and cyclin G1 also appear to be induced in the neurons bordering the ischemic core region after middle cerebral artery occlusion in rats.[31] The significance of the elevations in cell cycle

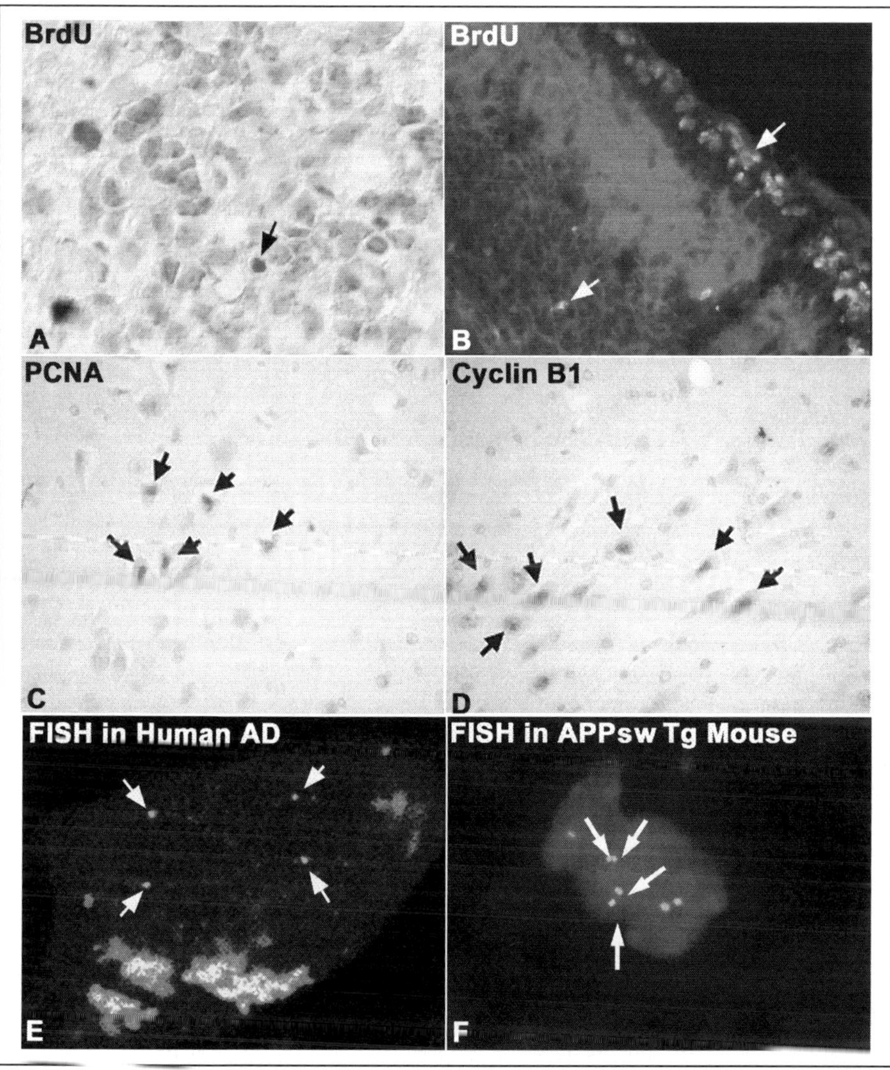

Figure 1. Immunocytochemical evidence for cell cycle events in models of neurodegeneration in mouse and man. BrdU is incorporated into post-mitotic cerebellar granule cells (arrows) if they are deprived of their target. Such a situation is found in mutations (A) such as staggerer and also during normal development in the wild type (B). Neurons in populations at risk for cell death in Alzheimer's disease (such as hippocampal pyramidal cells) also reexpress cell cycle markers such as PCNA (C) or cyclin B (D) before their death. These cell cycle proteins lead to a well regulated S-phase as illustrated by the incorporation of BrdU (A, B) or the use of fluorescent in situ hybridization to detect the duplication of specific loci in both human (E) and mouse (F). The probe shown in the human hippocampal neuron is a probe to a region of the human genome containing the end of the long arm of chromosome 11. The probe shown in F recognizes a long stretch of unique genomic sequences surrounding the aldolase C gene.

protein content is clear since administration of flavopiridol prevents neuronal death in the vulnerable CA1 region neurons,[32] and intraventricular administration of flavopiridol in the focal ischemia reduced the size of infarct.

Human Disease

The studies cited above provide solid evidence for the role of cell cycle events in the prosecution of neuronal cell death. What these studies also reveal is that a wide range of insults is able to induce a neuronal death that is associated with cell cycle reactivation. It seems only logical, therefore, that several neurodegenerative diseases have been found associated with an apparent induction of a cell cycle in the areas where neurons are lost.

Alzheimer's Disease

A key feature of the working hypothesis concerning the prohibition against a neuron reentering a cell cycle is that it applies from the moment a developing neuron leaves its neurogenic phase and persists until the death of the organism. This has led many laboratories to investigate the involvement of cell cycle events in a variety of neurodegenerative diseases. Of these, by far the best studied is sporadic Alzheimer's disease. Beginning with immunocytochemical evidence for the ectopic expression of a variety of cell cycle enzymes, nearly half a dozen laboratories have proposed that the neurons at risk for cell death in Alzheimer's disease reenter a cell division process before their death. The cell cycle proteins whose levels have been reported to increase include cyclin D,[33,34] cdk4,[33] PCNA,[33,35] cyclin B,[33] cdc2[36-38] and Ki67.[34,35] In addition to these proteins whose presence is usually found only in actively cycling cells a number of studies have pointed out that several of the Cdk inhibitors are also present. These include p16^{ink4},[39] p21^{ras}[39] and p105.[40]

As is often the case with post-mortem human studies the available tissue is derived predominantly from individuals who died with advanced, if not end stage disease. This leaves open the possibility of several types artifact. For example, it is possible that the observed cell cycle changes are a rare but stable event that collects throughout the 10-year disease course and the quantitative prevalence that is seen in end-stage material is not representative of the contribution of cell-cycle events to the overall disease pathology. Alternatively, it could be that cell cycling is only prevalent at the very end of the illness, once again leading to a false impression of its importance. To address this problem,[41] have done a detailed quantitative analysis of the prevalence of cell cycle events in individuals who die with a diagnosis of mild cognitive impairment (MCI). We adopted this approach because several studies have shown that a high percentage of individuals with MCI will progress to Alzheimer's disease (AD) within 3-5 years of diagnosis.[42] Thus many researchers view MCI as a prodromal stage of Alzheimer's disease.[43,44] Using a battery of cell cycle protein antibodies, we found that in both hippocampus and basal nucleus there is a significant percentage of cell cycle immunopositive neurons in all MCI cases (Fig. 1C, D). Indeed, the percentages turn out to be very similar to those found in Alzheimer's disease cases, and significantly higher than in cognitively intact controls. This means that cell cycle-related cell death is not a peculiarity of late stage disease and suggests that it represents a unifying disease mechanism.

An important question raised by these studies is whether or not the appearance of this large collection of proteins has any functional meaning in terms of actual cell cycle progression. To date, only a single study has addressed this question in the human. Yang et al[45] performed fluorescent in situ hybridization using either large genomic probes to unique sites in the human genome or small probes to the highly repetitive DNA of the centromere of specific chromosomes. The study showed that in two populations of nerve cells that are at risk for cell death in the CNS of the Alzheimer's disease brain (hippocampal pyramidal cells and the cholinergic neurons of the basal nucleus of Meynert) there were significant numbers neurons that showed evidence for three or four copies of each of these probes (Fig. 1E), direct indication that DNA replication has occurred. This means that the ectopic expression of cell cycle protein was sufficiently coordinated that a well-regulated S-phase ensued. Recent unpublished evidence from our lab suggests that the same process of cell cycle initiation occurs in engineered mouse models of AD (Fig. 1F). That said, recent data from[46] adds an unusual wrinkle to the story. In a comprehensive analysis of the identity of the DNA polymerases that are involved these authors found an induction of DNA polymerase-beta, typically associated with DNA repair, rather the replication polymerase, pol-delta.

Stroke and Other Human Diseases

As in the animal models of ischemia, postmortem studies of brain tissue in and around a stroke or ischemic event reveal evidence for the reexpression of cell cycle proteins such as PCNA, GADD34.[47,48] The timing of the appearance of these proteins strongly suggests that the initiation of a cell cycle is an early event in the process leading to cell death. Cell cycle markers have also been detected in brains of patients with Parkinson's disease,[49] in spinal cord samples from amyotrophic lateral sclerosis cases,[50] in frontotemporal dementia, in Neimann Picks disease, in Pick's disease, and in progressive supranuclear palsy.[51] In an experimental in vivo model of Parkinson's disease, dying neurons of the substantia nigra express factors specific for G2, S and M phase of cell cycle.[52] Activation of cell cycle machinery has also been observed in vulnerable regions following a traumatic brain injury[53] as well as several examples of infection induced neuronal loss, and other instances of neurodegenerative disease.[54-57] Finally, recent evidence implicates cell cycle anomalies in both amyotrophic lateral sclerosis[50,58] and the related SOD-1 transgenic mouse model.[59]

Conclusions

The cells of the adult central nervous system appear to be in a continual struggle to hold their cell cycle in check. This is evident both from the virtual absence of neuronal cell division in the adult and from the apparent fate of those neurons that do try to divide. The losses of nerve cells in a variety of neurodegenerative conditions, including several human diseases, seem to be related by this common theme of attempted neuronal cell division. At a superficial level, it is almost as if the neurons retain a developmental 'memory' of their epithelial origins and following stress or 'wounding' of the CNS, a suppressed urge to repair the 'wound' by cell division ensues. This attempt would seem laudable on its surface – in a situation where neurons are dying, any effort to create new nerve cells should be welcome. But these good intentions on the part of the mature neurons have unintended consequences, and in the final analysis end up paving a very famous road that leads only to a worsening of the brain's condition.

References

1. al-Ubaidi MR, Hollyfield JG, Overbeek PA et al. Photoreceptor degeneration induced by the expression of simian virus 40 large tumor antigen in the retina of transgenic mice. Proc Natl Acad Sci USA 1992; 89:1194-1198.
2. Hammang JP, Behringer RR, Baetge EE et al. Oncogene expression in retinal horizontal cells of transgenic mice results in a cascade of neurodegeneration. Neuron 1993; 10:1197-1209.
3. Feddersen R, Clark H, Yunis W et al. In vivo viability of postmitotitc Purkinje neurons requried pRb family member function. Mol Cell Neurosci 1995; 6:153-167.
4. DeCaprio JA, Ludlow JW, Figge J et al. SV40 large tumor antigen forms a specific complex with the product of the retinoblastoma susceptibility gene. Cell 1988; 54:275-283.
5. Harbour JW, Dean DC. Rb function in cell-cycle regulation and apoptosis. Nat Cell Biol 2000; 2:E65-67.
6. Tonini T, Hillson C, Claudio PP. Interview with the retinoblastoma family members: Do they help each other? J Cell Physiol 2002; 192:138-150.
7. Clarke A, Maandag E, van Roon M et al. Requirement for a functional Rb-1 gene in murine development. Nature 1992; 359:328-330.
8. Jacks T, Fazeli A, Schmitt E et al. Effects of an Rb mutation in the mouse. Nature 1992; 359:295-300.

9. Lee EY-HP, Chang C-Y, Hu N et al. Mice deficient for Rb are nonviable and show defects in neurogenesis and haematopoiesis. Nature 1992; 359:288-294.

10. Freeman R, Estus S, Johnson E. Analysis of cell cycle-related gene expression in postmitotic neurons: Selective induction of cyclin D1 during programmed cell death. Neuron 1994; 12:343-355.

11. Herrup K, Busser JC. The induction of multiple cell cycle events precedes target-related neuronal death. Development 1995; 121:2385-2395.

12. Ferrari G, Greene LA. Proliferative inhibition by dominant-negative Ras rescues naive and neuronally differentiated PC12 cells from apoptotic death. Embo J 1994; 13:5922-5928.

13. Ferrari G, Yan CY, Greene LA. N-acetylcysteine (D- and L-stereoisomers) prevents apoptotic death of neuronal cells. J Neurosci 1995; 15:2857-2866.

14. Mayo JC, Sainz RM, Uria H et al. Inhibition of cell proliferation: A mechanism likely to mediate the prevention of neuronal cell death by melatonin. J Pineal Res 1998; 25:12-18.

15. Park DS, Morris EJ, Stefanis L et al. Multiple pathways of neuronal death induced by DNA-damaging agents, NGF deprivation, and oxidative stress. J Neurosci 1998; 18:830-840.

16. Lambeng N, Michel PP, Brugg B et al. Mechanisms of apoptosis in PC12 cells irreversibly differentiated with nerve growth factor and cyclic AMP. Brain Res 1999; 821:60-68.

17. Farinelli S, Greene L. Cell cycle blockers mimosine, ciclopirox, and deferoxamine prevent the death of PC12 cells and postmitotic sympathetic neurons after removal of trophic support. J Neurosci 1996; 16:1150-1162.

18. Park D, Farinelli S, Greene L. Inhibitors of cyclin-dependent kinases promote survival of post-mitotic neuronally differentiated PC12 cells and sympathetic neurons. J Biol Chem 1996; 271:8161-8169.

19. Park D, Levine B, Ferrari G et al. Cyclin dependent kinase inhibitors and dominant negative cyclin dependent kinase 4 and 6 promote survival of NGF-deprived sympathetic neurons. J Neurosci 1997a; 17:8975-8983.

20. Park D, Morris E, Greene L et al. G1/S cell cycle blockers and inhibitors of cyclin-dependent kinases suppress camptothecin-induced neuronal apoptosis. J Neurosci 1997b; 17:1256-1270.

21. Park DS, Obeidat A, Giovanni A et al. Cell cycle regulators in neuronal death evoked by excitotoxic stress: Implications for neurodegeneration and its treatment. Neurobiol Aging 2000; 21:771-781.

22. Liu DX, Greene LA. Neuronal apoptosis at the G1/S cell cycle checkpoint. Cell Tissue Res 2001; 305:217-228.

22a. Chen P, Zindy F et al. Progressive hearing loss in mice lacking the cyclin-dependent kinase inhibitor Ink4d. Nat Cell Biol 2003; 5(5):422-426.

23. Caddy KW, Biscoe TJ. Structural and quantitative studies on the normal C3H and Lurcher mutant mouse. Philos Trans R Soc Lond B Biol Sci 1979; 287:167-201.

24. Herrup K, Mullen R. Regional variation and absence of large neurons in the cerebellum of the staggerer mouse. Brain Res 1979a; 172:1-12.

25. Herrup K, Mullen RJ. Staggerer chimeras: Intrinsic nature of Purkinje cell defects and implications for normal cerebellar development. Brain Res 1979b; 178:443-457.

26. Wetts R, Herrup K. Interaction of granule, Purkinje and inferior olivary neurons in lurcher chimaeric mice. I. Qualitative studies. J Embryol Exp Morphol 1982; 68:87-98.

27. Herrup K. Role of staggerer gene in determining cell number in cerebellar cortex. I. Granule cell death is an indirect consequence of staggerer gene action. Brain Res 1983; 313:267-274.

28. Sonmez E, Herrup K. Role of staggerer gene in determining cell number in cerebellar cortex. II. Granule cell death and persistence of the external granule cell layer in young mouse chimeras. Brain Res 1984; 314:271-283.

29. Guegan C, Levy V, David JP et al. c-Jun and cyclin D1 proteins as mediators of neuronal death after a focal ischaemic insult. Neuroreport 1997; 8:1003-1007.

30. Li Y, Chopp M, Powers C et al. Immunoreactivity of cyclin D1/cdk4 in neurons and oligodendrocytes after focal cerebral ischemia in rat. J Cereb Blood Flow Metab 1997; 17:846-856.

31. van Lookeren Campagne M, Gill R. Increased expression of cyclin G1 and p21WAF1/CIP1 in neurons following transient forebrain ischemia: Comparison with early DNA damage. J Neurosci Res 1998; 53:279-296.

32. Wang F, Corbett D, Osuga H et al. Inhibition of cyclin-dependent kinases improves CA1 neuronal survival and behavioral performance after global ischemia in the rat. J Cereb Blood Flow Metab 2002; 22:171-182.

33. Busser J, Geldmacher DS, Herrup K. Ectopic cell cycle proteins predict the sites of neuronal cell death in Alzheimer's disease brain. J Neurosci 1998; 18:2801-2807.

34. Smith MZ, Nagy Z, Esiri MM. Cell cycle-related protein expression in vascular dementia and Alzheimer's disease. Neurosci Lett 1999; 271:45-48.
35. Nagy Z, Esiri M, Cato A et al. Cell cycle markers in the hippocampus in Alzheimer's disease. Acta Neuropathol (Berl) 1997; 94:6-15.
36. Dranovsky A, Vincent I, Gregori L et al. Cdc2 phosphorylation of nucleolin demarcates mitotic stages and Alzheimer's disease pathology. Neurobiol Aging 2001; 22:517-528.
37. Pei JJ, Braak H, Gong CX et al. Up-regulation of cell division cycle (cdc) 2 kinase in neurons with early stage Alzheimer's disease neurofibrillary degeneration. Acta Neuropathol (Berl) 2002; 104:369-376.
38. Johansson A, Hampel H, Faltraco F et al. Increased frequency of a new polymorphism in the cell division cycle 2 (cdc2) gene in patients with Alzheimer's disease and frontotemporal dementia. Neurosci Lett 2003; 340:69-73.
39. Luth HJ, Holzer M, Gertz HJ et al. Aberrant expression of nNOS in pyramidal neurons in Alzheimer's disease is highly colocalized with p21ras and p16INK4a. Brain Res 2000; 852:45-55.
40. Masliah E, Mallory M, Alford M et al. Immunoreactivity of the nuclear antigen p105 is associated with plaques and tangles in Alzheimer's disease. Lab Invest 1993; 69:562-569.
41. Yang Y, Mufson EJ, Herrup K. Neuronal cell death is preceded by cell cycle events at all stages of Alzheimer's disease. J Neurosci 2003; 23:2557-2563.
42. Bennett DA, Wilson RS, Schneider JA et al. Natural history of mild cognitive impairment in older persons. Neurology 2002; 59:198-205.
43. Petersen RC. Aging, mild cognitive impairment, and Alzheimer's disease. Neurol Clin 2000a; 18:789-806.
44. Petersen RC. Mild cognitive impairment: Transition between aging and Alzheimer's disease. Neurologia 2000b; 15:93-101.
45. Yang Y, Geldmacher DS, Herrup K. DNA replication precedes neuronal cell death in Alzheimer's disease. J Neurosci 2001; 21:2661-2668.
46. Copani A, Sortino MA, Caricasole A et al. Erratic expression of DNA polymerases by beta-amyloid causes neuronal death. Faseb J 2002; 16:2006-2008.
47. Imai H, Harland J, McCulloch J et al. Specific expression of the cell cycle regulation proteins, GADD34 and PCNA, in the peri-infarct zone after focal cerebral ischaemia in the rat. Eur J Neurosci 2002; 15:1929-1936.
48. O'Hare M, Wang F, Park D. Cyclin-dependent kinases as potential targets to improve stroke outcome. Pharmacol Ther 2002; 93:135.
49. Jordan-Sciutto KL, Dorsey R, Chalovich EM et al. Expression patterns of retinoblastoma protein in Parkinson disease. J Neuropathol Exp Neurol 2003; 62:68-74.
50. Ranganathan S, Bowser R. Alterations in G(1) to S phase cell-cycle regulators during Amyotrophic lateral sclerosis. Am J Pathol 2003; 162:823-835.
51. Husseman JW, Nochlin D, Vincent I. Mitotic activation: A convergent mechanism for a cohort of neurodegenerative diseases. Neurobiol Aging 2000; 21:815-828.
52. El-Khodor B, Boksa P. Caesarean section birth produces long term changes in dopamine D1 receptors and in stress-induced regulation of D3 and D4 receptors in the rat brain. Neuropsychopharmacology 2001; 25:423-439.
53. Katano H, Masago A, Taki H et al. p53-independent transient p21(WAF1/CIP1) mRNA induction in the rat brain following experimental traumatic injury. Neuroreport 2000; 11:2073-2078.
54. Jordan-Sciutto KL, Wang G, Murphy-Corb M et al. Induction of cell-cycle regulators in simian immunodeficiency virus encephalitis. Am J Pathol 2000; 157:497-507.
55. Mandel S, Grunblatt E, Youdim M. cDNA microarray to study gene expression of dopaminergic neurodegeneration and neuroprotection in MPTP and 6-hydroxydopamine models: Implications for idiopathic Parkinson's disease. J Neural Transm Suppl 2000; 117-124.
56. Jordan-Sciutto KL, Murray Fenner BA, Wiley CA et al. Response of cell cycle proteins to neurotrophic factor and chemokine stimulation in human neuroglia. Exp Neurol 2001; 167:205-214.
57. Jordan-Sciutto KL, Wang G, Murphey-Corb M et al. Cell cycle proteins exhibit altered expression patterns in lentiviral-associated encephalitis. J Neurosci 2002; 22:2185-2195.
58. Ranganathan S, Scudiere S, Bowser R. Hyperphosphorylation of the retinoblastoma gene product and altered subcellular distribution of E2F-1 during Alzheimer's disease and amyotrophic lateral sclerosis. J Alzheimers Dis 2001; 3:377-385.
59. Nguyen MD, Boudreau M, Kriz J et al. Cell cycle regulators in the neuronal death pathway of amyotrophic lateral sclerosis caused by mutant superoxide dismutase 1. J Neurosci 2003; 23:2131-2140.

CHAPTER 3

The Role of Presenilins in the Cell Cycle and Apoptosis

Mervyn J. Monteiro

Abstract

Alzheimer's disease (AD) is the most common cause of dementia in the elderly, affecting approximately 10% of individuals by 65 years of age and 47% by 85 years of age. Whereas the majority of AD cases appear to be sporadic, and occur in individuals that have no apparent family history, a small percentage of cases (~5%), termed early-onset familial Alzheimer's disease (FAD) arise in individuals at an unusually young age, with some developing disease in as early as the third decade of life. Molecular genetic studies have revealed that the majority of FAD are associated with dominant inheritance of mutations in three genes, one encoding the β-amyloid precursor protein (APP) on chromosome 21, and two encoding homologous proteins presenilin 1 and 2 (PS1 and PS2) on chromosomes 14 and 1, respectively.[1] The cause of late-onset AD appears to be much more complex, as several genes have been implicated as modifiers or risk factors for the disease.[2] Although 100 years have elapsed since AD was first recognized as a separate disease entity, the last two decades have produced some of the most significant breakthroughs in our understanding of the disease. Despite this progress, the precise mechanisms that lead to the massive demise of neurons that characterize those afflicted with AD remain unresolved, and there is no known treatment to prevent or cure the disease. In this review, I will summarize information on the key proteins involved in AD pathology, especially as it relates to apoptosis, followed by a more in-depth focus on the role of presenilins in cell cycle regulation and apoptosis. I will describe why an understanding of the misregulation of cell cycle events and apoptosis in AD may provide valuable insights for possible therapeutic interventions to prevent or cure the disease.

Alzheimer's Disease Pathology

Although Alzheimer's disease (AD) can be clinically diagnosed with ~90% accuracy, the definitive test is the pathologic examination of brain at autopsy. Characteristic changes in the brain distinguish AD from other forms of dementia. Brains affected by AD contain a buildup of two lesions, neurofibrillary tangles (NFT) within nerve cells and extracellular plaques in areas important for memory and intellectual functions.[3] Much of the research into AD focuses on how these two lesions are formed and whether they are the primary cause, or a secondary manifestation of the disease. The formation of NFTs and plaques appears to be caused by an interplay between a number of factors, including age, genes, and the environment. NFTs are composed principally of paired helical filaments (PHFs), which are assembled from the

Cell Cycle Mechanisms and Neuronal Cell Death, edited by Agata Copani and Ferdinando Nicoletti. ©2005 Eurekah.com and Kluwer Academic / Plenum Publishers.

microtubule-associated tau protein.[4] Direct evidence that tau plays a role in neurodegeneration comes from molecular genetic studies of frontal-temporal dementia (FTD), a neurodegenerative disorder that shares several common features with AD: several different mutations have been mapped in both introns and exons of the tau gene are associated with FTD.[4,5]

The principal protein that is deposited to form senile plaques is a proteolytic fragment of the β-amyloid precursor protein (APP).[1,3] The normal function of APP, a single-pass transmembrane glycoprotein that localizes to the cell surface, is not fully understood. However, APP undergoes a complex pathway of protein trafficking and is proteolytically cleaved at multiple sites, producing various size fragments. One of these fragments, called amyloid-beta protein (Aβ), is the culprit that aggregates and forms the plaques that characterize AD. Aβ is generated by proteolytic cleavages at the extracellular domain by β-secretase and within the transmembrane domain by γ-secretase.[1] Most Aβ fragments consist of the first 40 residues of Aβ (Aβ40), although slightly longer fragments consisting of residues 1 through 42 and 1 through 43 (which are collectively referred to as Aβ42 for simplicity), which are both more fibrillogenic and amyloidogenic than Aβ, arise due to differences in cleavage at the COOH-terminal γ-site. The β-site APP-cleaving enzyme (BACE) was recently identified as the major β-secretase,[6] but the identity of γ-secretase is more controversial. Several studies have indicated that γ-secretase activity is, in large part, influenced by the expression of functionally active presenilin proteins, which has lead to the suggestion that presenilins are γ-secretase.[7,8] However, direct evidence that presenilins possess γ-secretase activity has not yet been demonstrated. Clearly, understanding the pathway and mode of Aβ production should yield clues about how Aβ fragments contribute to AD pathogenesis. However, at present, the precise mechanisms that regulate the production of Aβ fragments are not fully understood, although certain agents (such as those that induce cell stress and those that elevate Ca^{2+} levels in cells) increase their production.

Interestingly, cells that express presenilin genes with FAD mutations appear to alter the specificity and/or activities of γ-secretase, such that the generation and accumulation of the longer and more amyloidogenic Aβ forms are favored.[1,9] Other studies have indicated that the Aβ fragments may, in fact, be generated as cell undergo apoptosis,[10] and that certain FAD-associated mutations in APP genes may increase the susceptibility of APP proteins to be cleaved by certain caspases.[11] It is also possible that calcium misregulation and/or apoptosis may be involved in Aβ production in AD, especially because both of these processes are linked to altered properties of APP and presenilin genes with FAD mutations.[12,13] Apart from the role that amyloid may have in AD pathogenesis, there are many other hypotheses regarding the etiology of AD. Due to space restrictions, only one of these, the role of presenilins in cell cycle regulation and apoptosis, will be discussed further.

Presenilin Structure and Function

There are two presenilin genes (PS1 and PS2) in mammals, and the two proteins they encode share 67% identity. PS1 and PS2 mRNAs and proteins are widely expressed in a number of different human tissues, with high expression particularly noteworthy in human neurons.[9] There is some debate as to the exact topology and location of presenilin proteins in cells. However, most of the published data suggests that both proteins have eight transmembrane domains, that both proteins are localized to the endoplasmic reticulum (ER), where each is oriented with their N-terminal domains, a large loop spanning the sixth and seventh transmembrane domains, and C-terminal domains facing the cytoplasm.[9] Full-length PS1 and PS2 polypeptides are 467 and 448 amino acids in length, respectively. Intact polypeptides for both proteins have been detected, in many cases, in human brain and in cell extracts.[14-16] Quite frequently, however, the presenilin proteins are proteolytically cleaved by an unknown protease at a site located in the loop, generating two smaller fragments.[17] It has been suggested that the cleaved presenilin products are the true functional presenilin components,[18] however there is also good evidence to indicate that proteolytic cleavage is not required for presenilin activity.[15,19-21] The exact biologic function of presenilins in cells is not well understood, although the genes that encode them have been genetically and phenotypically linked to proteins involved in the Notch-signaling pathway and in apoptosis.[12,22,23] Molecular disruption of presenilin genes in mouse, *Drosophila*, and *C. elegans* results in abnormalities during embryo development. In mouse, disruption of PS1, but not PS2, is lethal. However, disruption of both PS1 and PS2 genes causes embryonic lethality at an even earlier stage than that produced by loss of PS1 alone, suggesting that the two proteins may have somewhat overlapping functions.[24,25] Although presenilins clearly play a role in early development, their roles in aging, and the mechanism(s) by which FAD mutations cause AD, is not well understood.

The Role of Presenilins in Apoptosis

A common theme that has emerged from studies of presenilin proteins is that they are involved in apoptosis. An initial indication of this role emerged from a screen of cDNAs that were able to inhibit receptor-induced apoptosis in T-cells.[26] Although the C-terminal fragment of PS2 was recovered in this screen, subsequent studies indicated that the fragment acted in a dominant negative manner, because overexpression of full-length PS2 increased apoptosis.[27] Several additional studies have indicated that overexpression of PS2 in both neuronal and dividing cells sensitizes the cells to apoptosis, and that expression of PS2 genes with FAD mutations are associated with even higher levels of apoptosis.[28-30] In many of these cells, it was necessary to challenge the cells with certain apoptotic stimuli, however in other cells, for example HeLa cells, these apoptotic properties were manifested in the absence of any additional apoptotic stimuli.[31]

The underlying mechanism for how PS2 induces apoptosis in these systems is being clarified. Results from several studies indicate that, as expected, the apoptosis induced by PS2 is associated with an increase in caspase activity, especially of caspase-3.[32,33] In addition, an increase in cytochrome C release from mitochondria, increased expression of the proapoptotic effector Bax, and lower expression of the anti-apoptotic protein Bcl-2 are seen during PS2 induction of apoptosis.[34]

More recently, PS2 overexpression was shown to increase the level and transcriptional activity of the tumor suppressor protein, p53.[29] It is well known that loss of functional p53 can lead to uncontrolled cellular proliferation, and that increases in p53 activity are induced by DNA damage and by certain forms of cellular stress that cause cell cycle arrest and/or apoptosis.[35] The mechanism by which PS2 overexpression increases p53 activity is not yet known. The level of p53 activity in cells is exquisitely regulated, by post-transcriptional mechanisms such as occlusion of the p53 transactivation domain, ubiquitination and degradation of the protein by the proteasome, and shuttling of the protein between the cytoplasm and the nucleus.[35,36] It will be interesting to determine which, if any, of these mechanisms are affected by PS2 overexpression.

In contrast to what is known about PS2 and apoptosis, there is some debate about the involvement of PS1 in apoptosis. Several studies suggest that PS1 is also involved in apoptosis,[37-41] but others suggests that it plays no such role.[42] The reason for this discrepancy is unclear but may be a product of comparisons in different cell types and/or the effects of different expression systems used in the studies. The studies that have indicated that PS1 is involved in apoptosis, in general, have reported that FAD mutants, but not wild type proteins, cause cells to be more sensitive to apoptotic stimuli.

This effect has been obtained in both primary and established neuronal cell lines, as well as a variety of nonneuronal cells. The mechanisms by which PS1 FAD mutants sensitize cells to apoptosis closely parallel those implicated by PS2, although minor differences are found. As in the case of PS2, overexpression of wild type PS1 and FAD-linked PS1 mutations in Jurkat and human embryonic kidney (HEK) 293 cells enhance Fas-mediated apoptosis, with FAD mutants exerting stronger enhancement of apoptosis compared to wild type PS1.[39,40] Closer examination of the putative pathway by which apoptosis was induced in the cells indicated that Jun kinase (JNK) was in fact inhibited by wild type PS1 but that FAD mutant PS1 proteins stimulated an increase in JNK activity.[39]

A different outcome was obtained by Kim et al[43] who found that overexpression of both wild type and FAD mutant PS1 proteins in B103 rat neuroblastoma cells inhibited H_2O_2-induced apoptosis, and that both wild type and mutant PS1 proteins suppressed JNK activity. In a different study, primary cultured hippocampal neurons and rat pheochromocytoma (PC12) cells were transiently transfected with adenoviral vectors expressing wild type and FAD mutant PS1 proteins, and apoptosis together with cell signaling via protein kinase B (PKB also called Akt) was studied.[44] PS1 proteins containing FAD mutations, but not wild type PS1 protein, enhanced apoptosis when overexpressed in either cell type. They further demonstrated that PS1FAD mutants decreased Akt activity and β-catenin levels in cells, while the activity of GSK-3β kinase was elevated.

These findings are in accord with the downstream signaling pathway of Akt. Akt activation increases phosphorylation of GSK-3β kinase, causing inhibition of the latter kinase.[45] This inhibition compromises the ability of GSK-3β kinase to phosphorylate β-catenin, and as a result of the lack of this modification, β-catenin is not targeted for degradation by the proteasome.[46] An accumulation of β-catenin in cells leads to the translocation of the protein into the nucleus, where, in combination with lymphoid-enhancing factor 1 and T cell transcription factors (LEF/TCF), it transactivates genes such as cyclin D and myc that are involved in cellular proliferation.[47] Therefore, as expected, inhibition of Akt by PS1 FAD mutants produced the opposite effect, decreasing β-catenin levels in cells. In fact, earlier evidence indicating that FAD PS1 mutants destabilize β-catenin was obtained by Zhang et al[48] who demonstrated by pulse-labeling studies that the turnover of β-catenin was dramatically increased in cells stably expressing FAD mutant PS1 proteins but not in cells expressing wild type PS1 proteins. The authors also demonstrated that loss of β-catenin in neurons leads to an increase in the cells' vulnerability to apoptosis induced by Aβ protein. Interestingly, the authors also found that β-catenin levels were reduced in brain lysates from FAD mutant PS1 carriers but not in brain lysates from patients with sporadic AD.[48]

Several subsequent studies have supported the idea that FAD PS1 and PS2 mutants cause increased turnover of β-catenin,[49,50] whereas another study suggested that PS1FAD mutations decrease β-catenin turnover.[51] The discrepancy in some of these reports could be due to technical difficulties in measuring β-catenin levels in cells, especially because β-catenin is found in two fractions, a soluble and E-cadherin (membrane)-bound fraction (see ref. 52). Importantly, loss of PS1 expression in cells leads to increased accumulation of β-catenin indicating that presenilin proteins do indeed play an important role in controlling the accumulation of this protein (see below).[53]

Although presenilin FAD mutants may destabilize β-catenin by inhibiting Akt and GSK-3β-mediated phosphorylation, as suggested above, the connection between PS FAD mutants and the turnover of β-catenin is likely to be more complicated because presenilin proteins and β-catenin are thought to bind to one another and form high molecular weight complexes.[54,55] A further complicating factor is that presenilin proteins are thought to bind to cell adhesion sites that contain membrane-associated β-catenin and E-cadherin.[56,57] Clearly, understanding the mechanism by which presenilin proteins modulate β-catenin degradation, and whether the modulation of β-catenin levels affects signaling or some other cell function may be particularly informative with respect to AD.

Presenilins Apoptosis and the Calcium Connection

Another well-studied aspect of presenilin biology is their involvement in calcium regulation.[13,58] Such a connection was first hypothesized after the discovery that presenilin proteins share weak structural homologies with Ca^{2+}-channels.[59] In addition, defects in calcium regulation were noticed in cells cultured from individuals harboring mutations in presenilin proteins, even before the presenilin genes were linked to AD.[60] The cultured presenilin FAD-mutant fibroblasts had elevated receptor-mediated calcium responses when challenged with agonists. Subsequent transgenic and knockout studies of presenilin genes have confirmed that this, and other aspects of calcium signaling, are perturbed in cells that bear FAD presenilin mutations. Overall, these studies have indicated that FAD presenilin mutations potentiate inositol 1,4,5-triphosphate (IP_3)-mediated calcium release,[37,61,62] and that presenilin proteins with FAD mutations appear to have defects in capacitative calcium entry (CCE), the mechanism by which depleted internal calcium stores are replenished.[63,64]

Additional circumstantially evidence linking presenilin proteins to calcium regulation comes from the demonstration that presenilin proteins bind three different calcium-binding proteins: calsenilin, calmyrin and sorcin.[65-67] Coexpression of calmyrin or calsenilin with presenilin proteins potentiates apoptosis induced by the presenilins.[66,68] Because calcium plays a central role in biologic processes, it is not surprising that proper regulation of calcium is crucial for cell survival.[69] The misregulation of calcium homeostasis induced by presenilin FAD mutations may be the underlying cause that sensitizes cells to apoptosis.[12] Consistent with this hypothesis, overexpression of a calbindin D28, a calcium-buffering protein suppressed apoptosis induced by presenilin genes.[70] Therefore, it will be important to confirm that calcium-signaling defects are the underlying cause by which FAD mutations in presenilin genes cause AD.

Presenilins and the Cell Cycle

The discovery that presenilin proteins are involved in apoptosis prompted us to examine if the proteins were also involved in regulation of the cell cycle, since these two processes appear to be regulated, in part, by common factors.[71,72] The fact that presenilin genes are ubiquitously expressed suggests that studies of the expression of the proteins in cells other than in neurons could be important. In earlier studies, we found that overexpression of PS2 in HeLa cells induced apoptosis, and that expression of the FADPS2 (N141I) mutation enhanced apoptosis.[31] By BrdU labeling studies, we found that overexpression of PS1 or PS2 proteins in dividing cells arrested cells in the G_1 phase of the cell cycle, and that FAD PS1and PS2 mutations potentiated the cell cycle arrest.[73,74] Similar BrdU labeling studies in HEK cells confirmed that the proteins induce cell cycle arrest.[75] Interestingly, a comparison of the levels of cell cycle arrest produced by overexpression of three different FAD PS1 mutants, which are associated with different ages of onset of AD,[74,76] showed a trend between age of onset and degree of cell-cycle arrest, with the most aggressive mutant (P117L, which causes disease as early as 27 years of age[77]) producing the highest degree of cell cycle arrest and the least aggressive mutant (E280A, which causes AD between the ages of 45-54 years) producing the lowest degree of arrest. Immunoblot analysis of proteins in these transfected cell lysates indicated that overexpression of wild type and mutant presenilin proteins were associated with decreased levels of β-catenin,[74] but not with substantial changes in the levels of other cell cycle-regulated proteins, such as, p53, c-Myc, Rb, p21, and p27.[73] Although it is not known if there is any relationship between the cell cycle alterations produced in these overexpression studies and AD, several aspects about presenilins, AD, and the cell cycle merit further discussion.

First, the modulation of β-catenin levels by presenilin proteins may be directly related to the cell cycle and apoptotic effects produced by presenilin proteins. β-catenin levels oscillate in dividing cells, peaking at the G_1/S phase boundary of the cell cycle.[78] In an elegant study, Orford et al[78] demonstrated that an elevation in β-catenin levels was crucial to drive cells from the G_1 phase of the cell cycle into S. Moreover, they found that an elevation in β-catenin levels produced by transgenic expression of the protein, acted as a potent inhibitor of apoptosis. Thus, the arrest of dividing cells at G_1 conferred by overexpression of presenilin proteins may be due to the increased instability of β-catenin and failure of the protein to reach a critical threshold necessary for enabling cells to cross from G_1 into S phase. By contrast, a decrease in β-catenin levels would be predicted to increase the susceptibility of the cells to apoptosis.

Second, the function of a protein can sometimes be revealed by where it localizes within cells. The presenilins have been localized to the ER and plasma membrane. However, Li et al[79] reported that in fibroblasts the proteins are also localized to centromeres and centrosomes. The authors proposed that because centromeres and centrosomes are important for chromosome segregation, FAD mutations in presenilins might increase chromosome missegregation leading to apoptosis.[79] This hypothesis is based on other studies that have demonstrated that chromosome missegregation and aberrant mitosis increase during aging.[80] Interestingly, chromosome missegregation and trisomy 21 mosaicism were indeed found to be elevated in fibroblasts cultured from AD patients.[81]

Third, there is also growing evidence that presenilins are directly involved in the control of cell proliferation. In contrast to the cell cycle arrest induced by overexpression of presenilins, hyperplasia and neoplasia have been associated with loss of presenilin expression in epithelial cells of transgenic mice.[53] As described previously, disruption of the murine PS1 gene (in PS1-/- knockout mice) causes death shortly after birth.[82,83] Interestingly, transgenic expression of human PS1 under the control of the Thy-1 promoter rescues PS1-/- mice, allowing them to survive to adulthood.[53] It appears that the mice survive by rescue of PS1 expression, driven by the Thy-1 promoter, in neurons, and possibly other cell types. However, the Thy-1 promoter fails to drive PS1 expression in skin, a tissue where PS1 is normally expressed. The consequence of the loss of PS1 expression in skin produces uncontrolled cellular proliferation, frequently resulting in neoplasia.[53] Moreover, cultured keratinocytes from the rescued mice have elevated levels of β-catenin and cyclin D1. These results suggest that presenilin proteins may function as a tumor suppressor protein in certain cell types: whereby loss of PS1 expression leads to uncontrolled growth, and overexpression of PS1 or PS2 induces cell cycle arrest.

Fourth, the level of presenilin expression in neuronal cells appears to regulate neuronal differentiation and neurogenesis.[84-88] Recent studies of wild type and mutant presenilin genes in mice have indicated that presenilin genes harboring FAD mutations, but not wild type proteins, are compromised in neurogenesis.[89] If FAD presenilin mutations in humans produce similar manifestations in neurogenesis, then dying neurons in brain may not be replenished, which might lead to AD.

The above results indicate that the levels of presenilin proteins must be tightly regulated in cells, especially because uncontrolled expression of the proteins produces dramatic changes in the phenotype or survival of cells. Interestingly, overexpression of PS1 in cells leads to the down-regulation of PS2 and vice versa, indicating that the two proteins are coordinately regulated by a set of limiting cellular factors.[90] The identification of factors that regulate presenilin levels is therefore of considerable interest. Two factors that regulate presenilin levels are p53 and ubiquilin.[91,92] Overexpression of p53 causes the down-regulation of PS1,[29,91] whereas overexpression of ubiquilin increases PS1 and PS2 levels.[92] Clearly it will be important to understand how presenilin protein levels are regulated in cells as this could lead to important insight into their role in cancer and AD.

Do Presenilin Genes Have Dual Functions in Cellular Proliferation and Aging?

To date, at least 75 different missense FAD mutations have been mapped in PS1 and at least 6 FAD mutations have been mapped in PS2. It is both curious and intriguing that the presenilin genes are "hot-spots" for mutations that cause AD given what is known about their functions. It is possible that these two genes have dual, perhaps even unrelated, functions, such as involvement in cell proliferation as well as in aging. A shift in the balance between these two states might lead to disease. A recent study of p53 illustrates the consequences of altering the balance in the activity of a key cellular protein. A mutant p53 protein was found to confer resistance to spontaneous tumors, but unexpectedly it also accelerated aging in mice.[93] This result indicates that cells contain regulatory proteins that serve dual roles, in mitogenic control and aging, and that a shift in the exquisite balance between these two states would accelerate one process over the other. In the case of presenilins, there is now good evidence that indicates that the level of presenilin expression can dictate whether a cell proliferates or dies: too little expression leads to proliferation in certain cells whereas too high expression causes cell cycle arrest or apoptosis. Because FAD mutations in PS1 and PS2 genes potentiate cell cycle arrest, we speculate that AD might be caused by an imbalance in mitogenic and aging-associated activities of presenilin proteins. Alternatively, FAD mutations in the presenilin genes could arise due to evolutionary pressures that select for proteins that confer resistance to certain forms of cancers (e.g., melanomas, given that overexpression of presenilin genes in keratinocytes leads to cell cycle arrest and loss of expression in keratinocytes causes hyperproliferation) but the detrimental consequence is they cause early-onset AD. Epidemologic studies that examine the incidence of certain cancers (e.g., melanomas) in families that carry FAD presenilin mutations could help to validate this hypothesis.

Apoptosis and Alzheimer's Disease

Although many studies have indicated that FAD mutations in presenilin genes increase the susceptibility of cells to apoptosis, the relevance of this phenomenon to AD pathogenesis in humans is as yet unknown, because most of studies have been performed in vitro. Besides the presenilins, there is growing evidence that also suggests that FAD mutations in APP also increase cell vulnerability to apoptosis.[11,94,95] These and other studies have lead to a growing appreciation that apoptosis may be a common mechanism associated with the loss of neuronal and nonneuronal cells not only in AD, but also in several different neurodegenerative disorders.[96-98] Considering that the time course of AD progression can vary from a few years to as many as 20 years (with an average duration of 10 years), the massive loss of neurons that is a characteristic of AD-afflicted brains is expected to proceed over an extended period of time. The estimated frequency of cell loss at any given time during AD is predicted to be approximately 1 to 4 per 10,000 cells.[99] Because apoptosis is a relatively rapid process (taking 8-24 hours), leaving behind little trace of its occurrence, the frequency of finding cells undergoing apoptosis in brain is expected to be quite rare. Careful examination of brains after autopsy for signs of apoptosis have begun to provide compelling evidence that apoptosis is increased in AD brains compared with nonafflicted control brains from individuals not afflicted with AD.[100] These results have indicated that several markers of apoptosis, such as nuclear DNA fragmentation,[101-104] activation of caspases,[105-108] and proteins involved in cell death cascades,[100,109] are elevated in AD brains, and that they occur at the expected frequency for AD progression. Still, additional studies are needed to determine the underlying causes that trigger these cells to succumb to this apoptotic fate. Studies are also needed to clarify whether patients with FAD mutations in presenilins and APP genes have increased apoptosis, as might be expected from the predictions obtained in cell culture.

Finally, other chapters in this book outline the involvement of cell cycle abnormalities in AD-afflicted brains. The challenge for the future will be to integrate the multitude of findings that are quickly emerging on AD into a coherent picture so that the etiology of AD becomes clear. Only then can rational therapies be devised to treat and/or cure this devastating and dehumanizing disease. Given the remarkable progress that has been achieved so far, it will not be long before this goal is reached.

Acknowledgments

Due to space restrictions it was not possible to discuss all of the important presenilin papers. I apologize to the authors for their omission, and encourage the readers to read the other relevant reviews on presenilins. MJM gratefully acknowledges the grant support from the National Institute on Aging (RO1AG16839).

References

1. Hardy J. Amyloid, the presenilins and Alzheimer's disease. Trends Neurosci 1997; 20:154-159.
2. Bertram L, Tanzi RE. Dancing in the dark? The status of late-onset Alzheimer's disease genetics. J Mol Neurosci 2001; 17:127-136.
3. Selkoe DJ. The cell biology of beta-amyloid precursor protein and presenilin in Alzheimer's disease. Trends Cell Biol 1998; 8:447-453.
4. Goedert M, Spillantini MG. Tau gene mutations and neurodegeneration. Biochem Soc Symp 2001; 67:59-71
5. Lee VM, Goedert M, Trojanowski JQ. Neurodegenerative tauopathies. Annu Rev Neurosci 2001, 24:1121-1159
6. Vassar R, Bennett BD, Babu-Khan S et al. Beta-secretase cleavage of Alzheimer's amyloid precursor protein by the transmembrane aspartic protease BACE. Science 1999; 286:735-741.
7. Selkoe DJ, Wolfe MS. In search of gamma-secretase: Presenilin at the cutting edge. Proc Natl Acad Sci USA 2000; 97:5690-5692.
8. Strooper BD, Annaert W. Presenilins and the intramembrane proteolysis of proteins: Facts and fiction. Nat Cell Biol 2001; 3:E221-225.
9. Haass C, De Strooper B. The presenilins in Alzheimer's disease—proteolysis holds the key. Science 1999; 286:916-919.
10. LeBlanc A. Increased production of 4 kDa amyloid beta peptide in serum deprived human primary neuron cultures: Possible involvement of apoptosis. J Neurosci 1995; 15:7837-7846.
11. Gervais FG, Xu D, Robertson GS et al. Involvement of caspases in proteolytic cleavage of Alzheimer's amyloid- beta precursor protein and amyloidogenic A beta peptide formation. Cell 1999; 97:395-406.
12. Mattson MP, Guo Q, Furukawa K et al. Presenilins, the endoplasmic reticulum, and neuronal apoptosis in Alzheimer's disease. J Neurochem 1998; 70:1-14.
13. Monteiro MJ, Stabler SM. In Calcium: The molecular basis of calcium action in biology and medicine. In: Pochet R, Donato R, Haiech J, Heizmann C, Gerke V, eds. Kluwer Academic Publishers, 2000:607-623.
14. Dewji NN, Do C, Singer SJ. On the spurious endoproteolytic processing of the presenilin proteins in cultured cells and tissues. Proc Natl Acad Sci USA 1997; 94:14031-14036.
15. Jacobsen H, Reinhardt D, Brockhaus M et al. The influence of endoproteolytic processing of familial Alzheimer's disease presenilin 2 on abeta42 amyloid peptide formation. J Biol Chem 1999; 274:35233-35239.
16. Culvenor JG, Evin G, Cooney MA et al. Presenilin 2 expression in neuronal cells: Induction during differentiation of embryonic carcinoma cells. Exp Cell Res 2000; 255:192-206.
17. Thinakaran G, Borchelt DR, Lee MK et al. Endoproteolysis of presenilin 1 and accumulation of processed derivatives in vivo. Neuron 1996; 17:181-190.
18. Levitan D, Lee J, Song L et al. PS1 N- and C-terminal fragments form a complex that functions in APP processing and Notch signaling. Proc Natl Acad Sci USA 2001; 98:12186-12190.
19. Steiner H, Romig H, Grim MG et al. The biological and pathological function of the presenilin-1 Deltaexon 9 mutation is independent of its defect to undergo proteolytic processing. J Biol Chem 1999; 274:7615-7618.

20. Steiner H, Romig H, Pesold B et al. Amyloidogenic function of the Alzheimer's disease-associated presenilin 1 in the absence of endoproteolysis. Biochemistry 1999; 38:14600-14605.

21. Morihara T, Katayama T, Sato N et al. Absence of endoproteolysis but no effects on amyloid beta production by alternative splicing forms of presenilin-1, which lack exon 8 and replace D257A. Brain Res Mol Brain Res 2000; 85:85-90.

22. Kopan R, Goate A. A common enzyme connects notch signaling and Alzheimer's disease. Genes Dev 2000; 14:2799-2806.

23. Fortini ME. Notch and presenilin: A proteolytic mechanism emerges. Curr Opin Cell Biol 2001; 13:627-634.

24. Donoviel DB, Hadjantonakis AK, Ikeda M et al. Mice lacking both presenilin genes exhibit early embryonic patterning defects. Genes Dev 1999; 13:2801-2810.

25. Herreman A, Hartmann D, Annaert W et al. Presenilin 2 deficiency causes a mild pulmonary phenotype and no changes in amyloid precursor protein processing but enhances the embryonic lethal phenotype of presenilin 1 deficiency. Proc Natl Acad Sci USA 1999; 96:11872-11877.

26. Vito P, Lacana E, D'Adamio L. Interfering with apoptosis: Ca(2+)-binding protein ALG-2 and Alzheimer's disease gene ALG-3. Science 1996; 271:521-525.

27. Vito P, Wolozin B, Ganjei JK et al. Requirement of the familial Alzheimer's disease gene PS2 for apoptosis. Opposing effect of ALG-3. J Biol Chem 1996; 271:31025-31028.

28. Deng G, Pike CJ, Cotman CW. Alzheimer-associated presenilin-2 confers increased sensitivity to apoptosis in PC12 cells. FEBS Lett 1996; 397:50-54.

29. Alves da Costa C, Paitel E, Mattson MP et al. Wild-type and mutated presenilins 2 trigger p53-dependent apoptosis and down-regulate presenilin 1 expression in HEK293 human cells and in murine neurons. Proc Natl Acad Sci USA 2002; 99:4043-4048.

30. Araki W, Yuasa K, Takeda S et al. Overexpression of presenilin-2 enhances apoptotic death of cultured cortical neurons. Ann N Y Acad Sci 2000; 920:241-244

31. Janicki S, Monteiro MJ. Increased apoptosis arising from increased expression of the Alzheimer's disease-associated presenilin-2 mutation (N141I). J Cell Biol 1997; 139:485-495.

32. Kim TW, Pettingell WH, Jung YK et al. Alternative cleavage of Alzheimer-associated presenilins during apoptosis by a caspase-3 family protease. Science 1997; 277:373-376.

33. Vito P, Ghayur T, D'Adamio L. Generation of anti-apoptotic presenilin-2 polypeptides by alternative transcription, proteolysis, and caspase-3 cleavage. J Biol Chem 1997; 272:28315-28320.

34. Araki W, Yuasa K, Takeda S et al. Pro-apoptotic effect of presenilin 2 (PS2) overexpression is associated with down-regulation of Bcl-2 in cultured neurons. J Neurochem 2001; 79:1161-1168.

35. Wahl GM, Carr AM. The evolution of diverse biological responses to DNA damage: Insights from yeast and p53. Nat Cell Biol 2001; 3:E277-286.

36. Vogelstein B, Lane D, Levine AJ. Surfing the p53 network. Nature 2000; 408:307-310.

37. Guo Q, Furukawa K, Sopher BL et al. Alzheimer's PS-1 mutation perturbs calcium homeostasis and sensitizes PC12 cells to death induced by amyloid beta-peptide. Neuroreport 1996; 8:379-383.

38. Guo Q, Sopher BL, Furukawa K et al. Alzheimer's presenilin mutation sensitizes neural cells to apoptosis induced by trophic factor withdrawal and amyloid beta-peptide: Involvement of calcium and oxyradicals. J Neurosci 1997; 17:4212-4222.

39. Wolozin B, Alexander P, Palacino J. Regulation of apoptosis by presenilin 1. Neurobiol Aging 1998; 19:S23-27.

40. Kirschenbaum F, Hsu SC, Cordell B et al. Substitution of a glycogen synthase kinase-3beta phosphorylation site in presenilin 1 separates presenilin function from beta-catenin signaling. J Biol Chem 2001; 276:7366-7375.

41. Hashimoto Y, Ito Y, Arakawa E et al. Neurotoxic mechanisms triggered by Alzheimer's disease-linked mutant M146L presenilin 1: Involvement of NO synthase via a novel pertussis toxin target. J Neurochem 2002; 80:426-437.

42. Bursztajn S, DeSouza R, McPhie DL et al. Overexpression in neurons of human presenilin-1 or a presenilin-1 familial Alzheimer disease mutant does not enhance apoptosis. J Neurosci 1998; 18:9790-9799.

43. Kim JW, Chang TS, Lee JE et al. Negative regulation of the SAPK/JNK signaling pathway by presenilin 1. J Cell Biol 2001; 153:457-463.

44. Nicholson KM, Anderson NG. The protein kinase B/Akt signalling pathway in human malignancy. Cell Signal 2002; 14:381-395.
45. Fukumoto S, Hsieh CM, Maemura K et al. Akt participation in the Wnt signaling pathway through Dishevelled. J Biol Chem 2001; 276:17479-17483.
46. Kim L, Kimmel AR. GSK3, a master switch regulating cell-fate specification and tumorigenesis. Curr Opin Genet Dev 2000; 10:508-514.
47. Morin PJ. β-catenin signaling and cancer. BioEssays 1999; 21:1021-1030
48. Zhang Z, Hartmann H, Do VM et al. Destabilization of beta-catenin by mutations in presenilin-1 potentiates neuronal apoptosis. Nature 1998; 395:698-702.
49. Nishimura M, Yu G, Levesque G et al. Presenilin mutations associated with Alzheimer disease cause defective intracellular trafficking of beta-catenin, a component of the presenilin protein complex. Nat Med 1999; 5:164-169.
50. Weihl CC, Miller RJ, Roos RP. The role of beta-catenin stability in mutant PS1-associated apoptosis. Neuroreport 1999; 10:2527-2532.
51. Kang DE, Soriano S, Frosch MP et al. Presenilin 1 facilitates the constitutive turnover of beta-catenin: Differential activity of Alzheimer's disease-linked PS1 mutants in the beta-catenin-signaling pathway. J Neurosci 1999; 19:4229-4237.
52. Orsulic S, Huber O, Aberle H et al. E-cadherin binding prevents beta-catenin nuclear localization and beta- catenin/LEF-1-mediated transactivation. J Cell Sci 1999; 112:1237-1245.
53. Xia X, Qian S, Soriano S et al. Loss of presenilin 1 is associated with enhanced beta-catenin signaling and skin tumorigenesis. Proc Natl Acad Sci USA 2001; 98:10863-10868.
54. Yu G, Chen F, Levesque G et al. The presenilin 1 protein is a component of a high molecular weight intracellular complex that contains beta-catenin. J Biol Chem 1998; 273:16470-16475.
55. Levesque G, Yu G, Nishimura M et al. Presenilins interact with armadillo proteins including neural-specific plakophilin-related protein and beta-catenin. J Neurochem 1999; 72:999-1008.
56. Georgakopoulos A, Marambaud P, Efthimiopoulos S et al. Presenilin-1 forms complexes with the cadherin/catenin cell-cell adhesion system and is recruited to intercellular and synaptic contacts. Mol Cell 1999; 4:893-902.
57. Singh N, Talalayeva Y, Tsiper M et al. The role of Alzheimer's disease-related presenilin 1 in intercellular adhesion. Exp Cell Res 2001; 263:1-13.
58. Mattson MP, LaFerla FM, Chan SL et al. Calcium signaling in the ER: Its role in neuronal plasticity and neurodegenerative disorders. Trends Neurosci 2000; 23:222-229.
59. Sherrington R, Rogaev EI, Liang Y et al. Cloning of a gene bearing missense mutations in early-onset familial Alzheimer's disease. Nature 1995; 375:754-760.
60. Ito E, Oka K, Etcheberrigaray R et al. Internal Ca2+ mobilization is altered in fibroblasts from patients with Alzheimer disease. Proc Natl Acad Sci USA 1994; 91:534-538.
61. Leissring MA, Paul BA, Parker I et al. Alzheimer's presenilin-1 mutation potentiates inositol 1,4,5-trisphosphate-mediated calcium signaling in Xenopus oocytes. J Neurochem 1999; 72:1061-1068.
62. Leissring MA, Parker I, LaFerla FM. Presenilin-2 mutations modulate amplitude and kinetics of inositol 1, 4,5-trisphosphate-mediated calcium signals. J Biol Chem 1999; 274:32535-32538.
63. Leissring MA, Akbari Y, Fanger CM et al. Capacitative calcium entry deficits and elevated luminal calcium content in mutant presenilin-1 knockin mice. J Cell Biol 2000; 149:793-798.
64. Yoo AS, Cheng I, Chung S et al. Presenilin-mediated modulation of capacitative calcium entry. Neuron 2000; 27:561-572.
65. Buxbaum JD, Choi EK, Luo Y et al. Calsenilin: A calcium-binding protein that interacts with the presenilins and regulates the levels of a presenilin fragment. Nat Med 1998; 4:1177-1181.
66. Stabler SM, Ostrowski LL, Janicki SM et al. A myristoylated calcium-binding protein that preferentially interacts with the Alzheimer's disease presenilin 2 protein. J Cell Biol 1999; 145:1277-1292.
67. Pack-Chung E, Meyers MB, Pettingell WP et al. Presenilin 2 interacts with sorcin, a modulator of the ryanodine receptor. J Biol Chem 2000; 275:14440-14445.
68. Lilliehook C, Chan S, Choi EK et al. Calsenilin enhances apoptosis by altering endoplasmic reticulum calcium signaling. Mol Cell Neurosci 2002; 19:552-559.
69. Berridge MJ, Lipp P, Bootman MD. The versatility and universality of calcium signalling. Nat Rev Mol Cell Biol 2000; 1:11-21.

70. Guo Q, Christakos S, Robinson N et al. Calbindin D28k blocks the proapoptotic actions of mutant presenilin 1: Reduced oxidative stress and preserved mitochondrial function. Proc Natl Acad Sci USA 1998; 95:3227-3232.

71. Meikrantz W, Schlegel R. Apoptosis and the cell cycle. J Cell Biochem 1995; 58:160-174.

72. Pucci B, Kasten M, Giordano A. Cell cycle and apoptosis. Neoplasia 2000; 2:291-299.

73. Janicki SM, Monteiro MJ. Presenilin overexpression arrests cells in the G1 phase of the cell cycle. Arrest potentiated by the Alzheimer's disease PS2(N141I)mutant. Am J Pathol 1999; 155:135-144.

74. Janicki SM, Stabler SM, Monteiro MJ. Familial Alzheimer's disease presenilin-1 mutants potentiate cell cycle arrest. Neurobiol Aging 2000; 21:829-836.

75. Jeong SJ, Kim HS, Chang KA et al. Subcellular localization of presenilins during mouse preimplantation development. Faseb J 2000; 14:2171-2176.

76. Van Broeckhoven C. Presenilins and Alzheimer disease. Nat Genet 1995; 11:230-232.

77. Wisniewski T, Dowjat WK, Buxbaum JD et al. A novel Polish presenilin-1 mutation (P117L) is associated with familial Alzheimer's disease and leads to death as early as the age of 28 years. Neuroreport 1998; 9:217-221.

78. Orford K, Orford CC, Byers SW. Exogenous expression of beta-catenin regulates contact inhibition, anchorage-independent growth, anoikis, and radiation-induced cell cycle arrest. J Cell Biol 1999; 146:855-868.

79. Li J, Xu M, Zhou H et al. Alzheimer presenilins in the nuclear membrane, interphase kinetochores, and centrosomes suggest a role in chromosome segregation. Cell 1997; 90:917-927.

80. Ly DH, Lockhart DJ, Lerner RA et al. Mitotic misregulation and human aging. Science 2000; 287:2486-2492.

81. Geller LN, Potter H. Chromosome missegregation and trisomy 21 mosaicism in Alzheimer's disease. Neurobiol Dis 1999; 6:167-179.

82. Wong PC, Zheng H, Chen H et al. Presenilin 1 is required for Notch1 and DII1 expression in the paraxial mesoderm. Nature 1997; 387:288-292.

83. Shen J, Bronson RT, Chen DF et al. Skeletal and CNS defects in Presenilin-1-deficient mice. Cell 1997; 89:629-639.

84. Hong CS, Caromile L, Nomata Y et al. Contrasting role of presenilin-1 and presenilin-2 in neuronal differentiation in vitro. J Neurosci 1999; 19:637-643.

85. Handler M, Yang X, Shen J. Presenilin-1 regulates neuronal differentiation during neurogenesis. Development 2000; 127:2593-2606.

86. Paganelli AR, Ocana OH, Prat MI et al. The Alzheimer-related gene presenilin-1 facilitates sonic hedgehog expression in Xenopus primary neurogenesis. Mech Dev 2001; 107:119-131.

87. Yang X, Handler M, Shen J. Role of presenilin-1 in murine neural development. Ann N Y Acad Sci 2000; 920:165-170

88. Wen PH, Friedrich Jr VL, Shioi J et al. Presenilin-1 is expressed in neural progenitor cells in the hippocampus of adult mice. Neurosci Lett 2002; 318:53-56.

89. Wen PH, Shao X, Shao Z et al. Overexpression of wild type but not an FAD mutant presenilin-1 promotes neurogenesis in the hippocampus of adult mice. Neurobiol Dis 2002; 10:8-19.

90. Thinakaran G, Harris CL, Ratovitski T et al. Evidence that levels of presenilins (PS1 and PS2) are coordinately regulated by competition for limiting cellular factors. J Biol Chem 1997; 272:28415-28422.

91. Roperch JP, Alvaro V, Prieur S et al. Inhibition of presenilin 1 expression is promoted by p53 and p21WAF-1 and results in apoptosis and tumor suppression. Nat Med 1998; 4:835-838.

92. Mah AL, Perry G, Smith MA et al. Identification of ubiquilin, a novel presenilin interactor that increases presenilin protein accumulation. J Cell Biol 2000; 151:847-862.

93. Tyner SD, Venkatachalam S, Choi J et al. p53 mutant mice that display early ageing-associated phenotypes. Nature 2002; 415:45-53.

94. Nishimoto I. A new paradigm for neurotoxicity by FAD mutants of betaAPP: A signaling abnormality. Neurobiol Aging 1998; 19:S33-38.

95. Yamatsuji T, Matsui T, Okamoto T et al. G protein-mediated neuronal DNA fragmentation induced by familial Alzheimer's disease-associated mutants of APP. Science 1996; 272:1349-1352.

96. Cotman CW. Apoptosis decision cascades and neuronal degeneration in Alzheimer's disease. Neurobiol Aging 1998; 19:S29-32.

97. Mattson MP. Apoptosis in neurodegenerative disorders. Nat Rev Mol Cell Biol 2000; 1:120-129.

98. Shimohama S. Apoptosis in Alzheimer's disease—An update. Apoptosis 2000; 5:9-16.

99. Perry G, Nunomura A, Smith MA. A suicide note from Alzheimer disease neurons? Nat Med 1998; 4:897-898.

100. Guo Q, Fu W, Xie J et al. Par-4 is a mediator of neuronal degeneration associated with the pathogenesis of Alzheimer disease. Nat Med 1998; 4:957-962.

101. Su JH, Anderson AJ, Cummings BJ et al. Immunohistochemical evidence for apoptosis in Alzheimer's disease. Neuroreport 1994; 5:2529-2533.

102. Sugaya K, Reeves M, McKinney M. Topographic associations between DNA fragmentation and Alzheimer's disease neuropathology in the hippocampus. Neurochem Int 1997; 31:275-281.

103. Adamec E, Vonsattel JP, Nixon RA. DNA strand breaks in Alzheimer's disease. Brain Res 1999; 849:67-77.

104. Anderson AJ, Stoltzner S, Lai F et al. Morphological and biochemical assessment of DNA damage and apoptosis in Down syndrome and Alzheimer disease, and effect of postmortem tissue archival on TUNEL. Neurobiol Aging 2000; 21:511-524.

105. Stadelmann C, Deckwerth TL, Srinivasan A et al. Activation of caspase-3 in single neurons and autophagic granules of granulovacuolar degeneration in Alzheimer's disease. Evidence for apoptotic cell death. Am J Pathol 1999; 155:1459-1466.

106. Rossiter JP, Anderson LL, Yang F et al. Caspase-cleaved actin (fractin) immunolabelling of Hirano bodies. Neuropathol Appl Neurobiol 2000; 26:342-346.

107. Su JH, Nichol KE, Sitch T et al. DNA damage and activated caspase-3 expression in neurons and astrocytes: Evidence for apoptosis in frontotemporal dementia. Exp Neurol 2000; 163:9-19.

108. Rohn TT, Head E, Nesse WH et al. Activation of caspase 8 in the Alzheimer's disease brain. Neurobiol Dis 2001; 8:1006-1016.

109. Su JH, Deng G, Cotman CW. Bax protein expression is increased in Alzheimer's brain: Correlations with DNA damage, Bcl-2 expression, and brain pathology. J Neuropathol Exp Neurol 1997; 56:86-93.

CHAPTER 4

Cell Cycle Activation and Cell Death in the Nervous System

Zsuzsanna Nagy

Abstract

The discovery of the cell division cycle opened new avenues for the understanding of cancer as well as in the search for therapy. However, the implications of the discovery had a more profound effect on biological research than anticipated at the time. We can now clearly distinguish between the activation of the cell division cycle and cell division itself. We also have closer understanding of the differences between senescence, quiescence and terminal differentiation. Furthermore the elucidation of the mechanisms that regulate the cell cycle also means that we have now begun to understand the mechanisms that link cell division and cell death.

Introduction

Classically the cell cycle is regarded as the pathway leading to cellular proliferation. In most cells, both in vivo and in vitro, the activation of the cell cycle by mitogenic factors leads to DNA replication (S phase) and to mitosis (M phase) (Fig. 1). The unhindered progression of the cell through the phases of cell cycle is tightly controlled by the sequential activation of cyclin/cyclin dependent kinase complexes. This is dependent upon the sequential expression of the catalytic cyclin subunits (Fig. 2) and the expression of cyclin dependent kinase inhibitors (CDKI – Fig. 1) (reviewed in ref. 1). The expression of CDKIs is triggered in the presence of differentiation factors that are cell-type dependent. DNA damage can also lead to the expression of these molecules, leading to cell cycle arrest to allow for DNA repair or cell death (Fig. 1).

Apoptosis is a highly conserved mechanism of active cell death that is involved in the development of tissues and organs. The phenomenon was first recognised, as distinct from necrosis by its morphological characteristics (reviewed in ref. 2). It emerged later that this cell suicide programme is more that just a developmentally regulated event, shaping embryonic organs, but it plays a vital role in the maintenance of tissue homeostasis in the adult organism. Furthermore, it appears that it is this mechanism that plays a pivotal role in many disease conditions, making it one of the most desirable targets for therapy. The molecular regulators of apoptosis are members of the bcl-2 protein family. Some members of the family (such as bcl-2, bcl_{-xL}) suppress apoptosis while others (Bax, Bak and Bik) are able to induce it. Since different members of the family form hetero- and homodimers inhibiting one other the fate of a cell will depend on the ratio between these proteins. The downstream executioners of the apoptotic

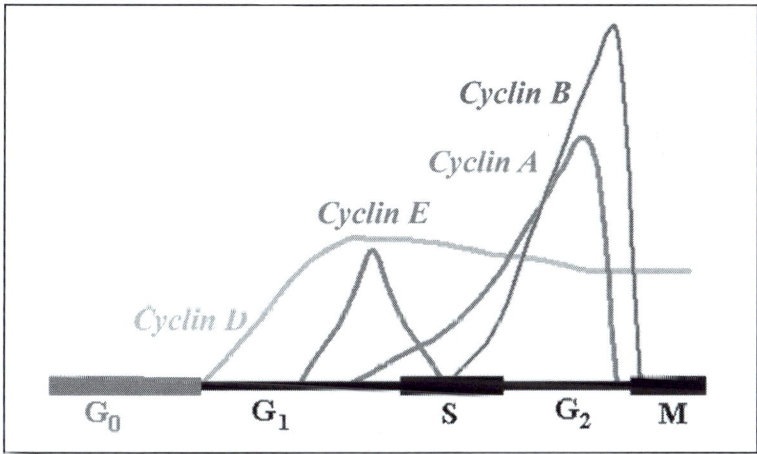

Figure 1. The regulation of the cell division cycle. G0= resting or quiescent phase; G1= fist gap phase; S= DNA replication phase; G2= second gap phase; M= mitosis. Cy= cyclin; CDK= cyclin dependent kinase. Arrows= initiation; flat arrows= inhibition.

Figure 2. The expression pattern of the cyclins during the cell division cycle.

process are the caspases, a set of enzymes that function together to form proteolytic cascades that are capable of autoactivation and ultimately lead to the breakdown of the cellular structure. The chain of events that link the two groups of molecules, the regulators and executioners, is still matter of debate.[3]

The links between cell division and apoptotic cell death have been extensively documented originally in cancer and in developmental studies. It became evident however, that this relationship is equally important in degenerative diseases of the central nervous system.

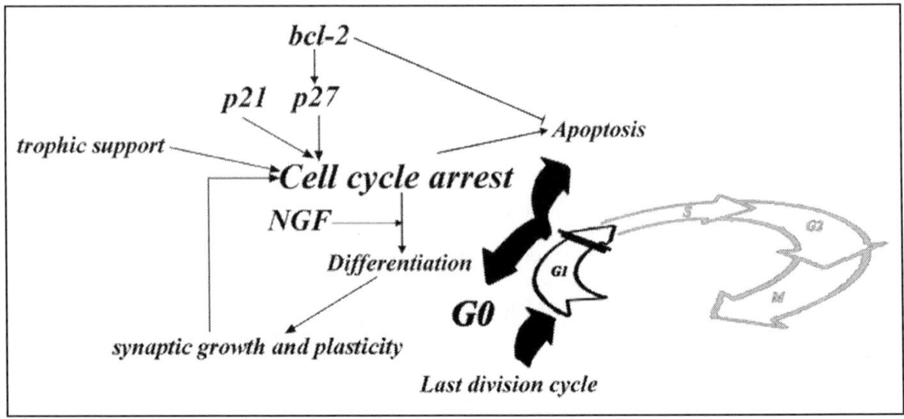

Figure 3. Apoptosis and the cell division cycle during development.

Cell Cycle and Apoptosis in the Developing Nervous System

Although the death of a large proportion of neurones in the developing brain may appear wasteful, programmed cell death (apoptosis) is a necessary part of brain patterning, just as important as the preceding division of the neuronal progenitors, the waves of migration, differentiation and synaptogenesis. The interplay between these cellular phenomena is complex and far from elucidated.[4]

During the development of the central nervous system the arrest of the last cell division cycle coincides with the beginning of terminal differentiation of neurones (Fig. 3, reviewed in refs. 5,6). The absence of cell cycle inhibitory cues lead to the lack of programmed cell death and results in a highly disorganized brain structure[7,8] while the forced and untimely arrest of the cell division process in the developing nervous system induces massive cell death and severe dysfunction.[9,10] These findings indicate that the negative regulation of cell division in neuronal precursors represents the first step towards the final brain patterning and it is necessary for the differentiation and survival of specific populations of central nervous system neurons.[11-13]

The induction of cell cycle arrest in the G1 phase in neuroblasts leads to increased sensitivity and responsiveness towards differentiation factors.[12] The induction of p21cip1, and cyclin dependent kinase inhibitor, is necessary for NGF dependent differentiation.[12,14,15] However, p21cip1 on its own it is not sufficient to induce neuronal differentiation. Additionally, the expression of proteins necessary for synaptic growth and plasticity in adulthood plays an important role in cell cycle arrest and differentiation of neuronal progenitors.[16] Furthermore, the well-documented anti-apoptotic function of the bcl-2 protein,[17,18] which is also involved in neuronal differentiation and axonal growth and regeneration,[19-21] is associated with a cell cycle inhibitory effect in dividing cell populations. The cell cycle control function of the bcl-2 protein is related to its ability to induce G1 arrest via the p27kip1 pathway.[22] Interestingly the withdrawal of trophic support from differentiated neurones leads to the reactivation of cell cycle and subsequent apoptotic death. These cells are efficiently rescued by the expression of CDKIs.[15,23]

It is also apparent that in the developing nervous system the apoptotic pathways inducible in resting or actively proliferating neuronal cells appears to be different, requiring the activation of different regulatory pathways.[24,25]

While the involvement of cell cycle arrest in apoptosis in the developing brain is well established the role of these phenomena in the adult brain are still debated. There is more and more emerging evidence however, that conditions affecting cell survival may initiate the reentry of neurones into the cell division cycle and result in subsequent cell death.

Iatrogenic Neurodegeneration

The vulnerability of neuronal precursor cells to cancer therapy is a long known phenomenon. It is also becoming more widely recognised that cancer therapies affect the adult central nervous system as a whole often leading to cognitive impairment in patients treated for both CNS and nonCNS cancers.[26-29] This deleterious effect[29,30] might become the dose-limiting factor in their use for cancer therapy.

Similar to the developing brain,[31] neuronal precursors in the dentate gyrus[32] and in the subeppendimal layer[33] of adult animals shows increased apoptosis after radiation. Although gamma irradiation induced neuronal cell death does not seem to require the presence of p53, neurogenesis is delayed in the presence of this protein after DNA damage,[34] indicating that irradiation-induced death and reduced proliferation of neuronal progenitors are closely linked phenomena. In contrast to this, the apoptotic death of differentiated neurones induced by ionising radiation does not seem to be mediated by cell cycle reentry and subsequent mitotic catastrophe but is associated with persistent DNA damage and activation of caspases.[35] It is a promising discovery that radiation-induced neuronal death can be prevented by tetracycline derivatives.[36] The precise mechanism of action by which these drugs exert their neuroprotective effects is not known, but they seem to affect both cell proliferation and survival.[37]

In contrast to ionising radiation other DNA damaging agents, such as UV radiation and cytotoxic drugs, induce neuronal death via the activation of the cell division cycle[38] and the cell death follows a mitotic catastrophe that can be prevented by cyclin dependent kinase inhibitors.[15,38]

The mechanism by which cytotoxic drugs may effect cognition are not documented by post-mortem human studies, but recent evidence indicates that some cytotoxic drugs, such as cisplatin, induce cell cycle arrest and apoptosis in dorsal root ganglion cells in vivo.[39]

Interestingly while cyclophosphamide seems to affect the healthy nervous system adversely there are studies indicating that after injury, ischaemic or mechanical, the use of cyclophosphamide actually prevents delayed neuronal damage.[40,41]

Neurodegenerative Disorders

The discovery of cell cycle related events in the pathogenesis of chronic neurodegenerative disorders opened new lines of enquiry in neuroscience.[42] There is still much debate and controversy, but the number of studies examining the role of cell cycle in the formation of specific pathologies and related cell death in the brain is increasing.

Alzheimer's Disease and Related Dementias

The search for the cellular processes underlying the formation of Alzheimer-type pathology took an unexpected turn, when cell cycle regulatory proteins have been found to be involved.[43-47]

The initial finding of nuclear Ki 67, cyclin E and cyclin B indicated that neurones may reenter the cell division cycle even if they are not capable of actual mitosis.[44,48,49] The general distribution of the cell cycle related protein expression and its relationship to either tau phosphorylation or apoptosis related proteins[50,51] lead to the formulation of the cell cycle hypothesis for the pathogenesis of AD.[52] This hypothesis postulates that neurones in the adult central nervous system are not terminally differentiated and they are capable of reentering the cell division cycle.[52] The cell cycle reentry on its own is not necessarily a harmful process, since the early arrest of the cell division cycle might be followed by redifferentiation without any functional consequences.[52] However, the progression of the cell division cycle into later phases will necessarily mean that differentiation is not possible any longer. Cells arrested in the late G2 phase of the cell cycle may either undergo programmed cell death or, if they survive, they will produce the typical Alzheimer's type pathology.[52] The hypothesis, as formulated at time, posed several questions. What are the factors that act as mitogenic stimuli for the neuronal populations affected in AD? Why is cell cycle progression allowed into the late, G2 phase? What is the exact chain of events that occurs from the moment of mitotic stimulation in the affected neurones? Why do some neurones develop AD-type pathology as a result of cell cycle progression while some others just die?

Subsequently several cell cycle related proteins were studied in relation to AD-type pathology and neuronal death[43,51,53-63] in an attempt to understand the process.

Neuronal populations vulnerable to the development of Alzheimer-type pathology are those that retain a high synaptic plasticity in the adult brain.[64-66] These neurones retain immature features and are not fully differentiated. This in turn allows their reentry into the cell division cycle if differentiation control fails.[65] This hypothesis is well supported by the observed pattern of synaptic remodelling in the adult brain and the expression of cell cycle-related proteins in Alzheimer's disease (AD).[52,66,67] It appears therefore that the high plasticity of these neuronal populations is a necessary prerequisite for cell cycle reentry in neurones. Therefore, genotypic or environmental factors affecting synaptic plasticity will inadvertently have an effect on the development of Alzheimer's disease. It has been shown that the ApoE4 genotype, which is a risk factor for Alzheimer's disease, is indeed associated with reduced neuroplastic capacity of neurones.[68] The activity dependent nature of self-organisation of the brain structure may explain the link between low education and an increased prevalence of Alzheimer's disease. Additionally the known differentiating effects of estrogens and thyroid hormone and their contribution to synaptic stability may explain the protective effect of these hormones against AD.[69,70] However, since cell cycle reentry on its own is not enough to cause Alzheimer's disease, morphodysgenesis as the trigger for cell cycle reentry,[65] appears to be a only a prerequisite which on its own is not sufficient to cause the disease.

The difference between the healthy aging and AD is the extent of cell cycle progression rather than cell cycle reentry.[52,71] Our initial findings support this view, together with other studies, indicating that markers of the late phases of the cell cycle[46,49] and DNA replication[55] precede the development of AD-type pathology. This suggests that the leap from healthy ageing to AD is the moment when neurones bypass the G1/S restriction point pointing to a possible regulatory fault at this checkpoint. The expression pattern of cell cycle regulator proteins of the G1/S transition has been examined in detail. Various studies indicate that these proteins are expressed in neurones in AD brains and are mostly associated with AD-type pathology.[43,60] The development of AD-type amyloid and tau pathology were linked to the late G2 phase of the cell cycle.[72-74] Therefore the finding of G1 inhibitor CDKIs associated with AD-type neuritic pathology in itself suggests that although these inhibitors are expressed in neurones, they are not sufficient to arrest the progression of the cell cycle. Another aspect of this association may be that the expression of these inhibitors rescues neurones from cell death allowing time for the development of AD-type pathology. In this context the accumulation of AD-type phospho tau may be a sign of neuroprototective mechanisms that prevent the rapid apoptotic death of neurones that have reentered the cell cycle.[64] This is further supported by the findings that the number of neurones that actually show signs of active apoptosis in AD is very small[8,75,76] despite of extensive DNA fragmentation. More detailed studies indicate that neurones either do not posses or do not activate the caspases responsible for the execution of the apoptotic cascade. This would explain the survival of neurones that have progressed in late phases of the cell cycle and the chronic long-lasting nature of AD.[76,77] Additionally it further justifies the discrimination between apoptosis and of the slow agonising death of neurones,

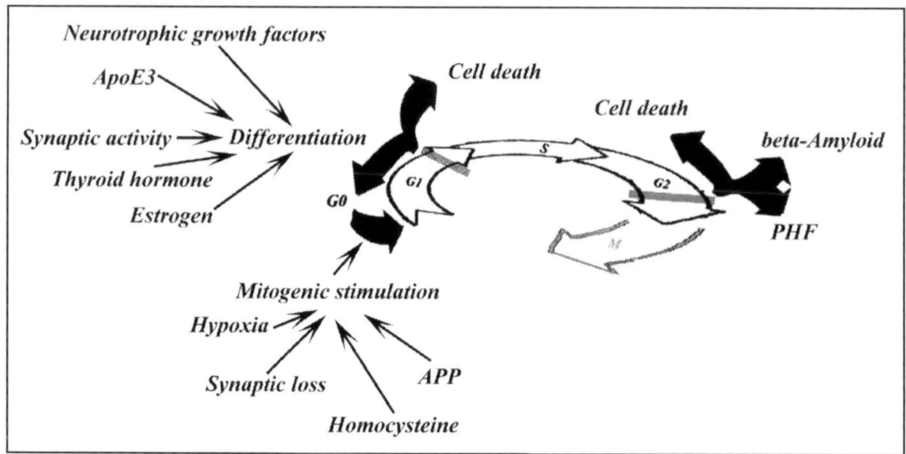

Figure 4. The neuronal division cycle in Alzheimer's disease.

aposklesis,[78] seen in neurodegenerative diseases. It is another question how long can these 'protective' mechanisms prevent neuronal death.

The finding of DNA replication[55] in AD neurones raises further issues. The 'comfort' of our original hypothesis postulating that the arrest of the cell cycle at the end of G2 in neurones might be due to the lack of DNA replication is lost, and the question of why neurones do not go though mitosis proper remains to be answered. One proposed explanation is that the cyclin dependent kinases are not translocated into the nucleus to drive the division cycle beyond the G2/M boundary.[42] Contrary to this hypothesis we find that even when the G2-related proteins are expressed in the nucleus mitosis does not occur. It is possible to speculate therefore, that the G2 arrest is due to incomplete DNA replication or extensive DNA damage in these cells. While the former cannot be answered with certainty at the moment the latter possibility is supported by growing evidence that links AD and oxidative DNA damage.[79]

Most of the initial cell cycle research has been done on sporadic cases, but there is growing evidence that this pathway is not restricted to this form of Alzheimer's disease. Presenilins (that play a pivotal role in the development of FAD in a small number of cases) also effects neuronal proliferation and survival.[80,81] It is believed that beta-amyloid itself may play a role in initiating cell cycle reentry and contributing to neuronal pathology in families carrying the APP mutations.[82-85] Additionally microglia activated by the aggregated beta-amyloid may release potentially mitogenic factors that contribute to the mitogenic reentry of neurones[86] (Fig. 4).

There is also evidence that ischaemia induced delayed neuronal death is due to the activation of the cell cycle, mediated by the loss of CDKI function in vivo.[87] Animal studies indicate that cyclin D expression precedes experimentally induced ischaemic neuronal death.[88-92] Similarly the phosphorylation of the retinoblastoma protein is altered by transient brain ischaemia probably representing a further step in the mitogenic signalling towards cell death.[93] In vitro studies also confirm that sublethal hypoxia may promote cell division-like phenomena.[94,95] While in acute hypoxic stress the activation of cell cycle may trigger neuronal death,[96] in case of sublethal stimulus the rapid expression of cell cycle inhibitors, such as p53 and p21cip1, may prevent cell death.[97]

The finding of a common pathogenic pathway for Alzheimer's disease and cerebro-vascular disease is not entirely unexpected. It is well known that the two diseases very often (about 20% of all demented patients) occur together. In patients where both pathologies are present the amount of Alzheimer-type pathology is significantly reduced relative to those without any

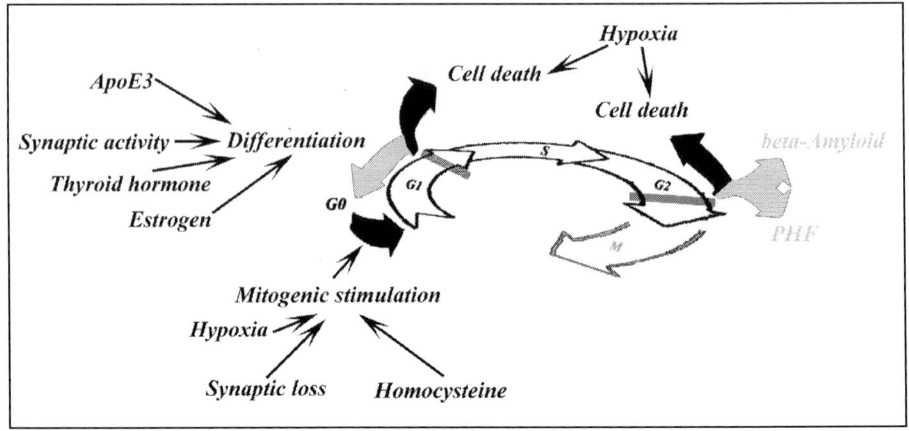

Figure 5. The neuronal cell cycle in cerebrovascular disease.

vascular component,[98,99] indicating that the AD-type degeneration and ischaemia-induced cell death represent alternative outcomes of the same cellular process[100,101] (Fig. 5).

Temporal Lobe Epilepsy

The massive cellular fallout in temporal lobe epilepsy is a long recognised consequence of the sustained seizure activity.[102] While the consequences of neuronal death and glial scaring have been assessed extensively little was known about the mechanisms that lead to seizure-related neuronal death. Less than a decade ago animal studies indicated that neuronal death associated with status epilepticus is due to apoptosis,[103] however it took another year until apoptosis, as a means of neuronal death in the adult nervous system, has been proved.[104]

Studies in human biopsy material indicate that the neuronal death in temporal lobe epilepsy is also associated with the expression of cell cycle-related proteins.[48,96] Seizure activity itself is capable of triggering cell cycle reentry in neurones[105] and promotes the progression of the cell cycle into its late G2 phase.[48] Moreover the seizure activity increases the ratio of Bax/Bcl-2 favouring apoptosis rather than redifferentiation or survival of the neurones.[106] In animal models of epilepsy exitotoxic stimuli induce cell cycle reactivation and cell death.[96,105,107]

In patients where epilepsy is successfully controlled by drugs the number of neurones that express the G2 specific cyclin B are negligible although the early phase cyclin E is present in numbers comparable to that found in patients suffering from intractable temporal lobe epilepsy.[48] This is hardly surprising if we take into account the anti-apoptotic effect of the anti-epileptic drugs that elevate the expression of bcl-2.[108]

It is also interesting that electroconvulsive therapy (ECT) may also trigger cell cycle reentry in certain vulnerable brain regions.[109] The possible interpretation of these findings however, is not equivocal. Whether the ECT indeed induces neurogenesis or it contributes to neuronal dysfunction and possibly death[110] via a cell cycle related pathway is another issue that remains to be explained.

In summary it appears that the reactivation of cell division like phenomena is a common pathway that leads to neuronal death (apoptosis or aposklesis) or specific cellular pathology in various neurodegenerative conditions. Although cell cycle re-entry is a common neurodegenerative pathway the outcome of the process varies depending on the genetic and epigenetic factors involved. As we unravel these determining factors it is becoming realistic to envisage that the discovery of links between the cell cycle and cell death in neurodegenerative disorders might hold the key to the future of an early diagnostic method[111-113] and treatment.

A Few Words of Caution

The reader might have noticed the omission of cdk-5 from this review despite its role in central nervous system patterning and in neurodegenerative conditions, such as Alzheimer's disease. This is not to diminish the role of this kinase in physiological and pathological conditions. Far from it. The scope of this chapter was to review the link between the activation of the cell division cycle and cell death in the nervous system. Despite its very suggestive name that leads to confusion sometimes, cdk-5 is not a cell cycle-related protein. The name is due to its homology to cdk-1. However, it was quickly identified as not playing a role in cell cycle control (reviewed in ref. 114). Therefore, without wishing to dismiss the role of cdk-5 in development or neurodegeneration, I had to refrain from discussing this molecule in order to fit in with the scope of this chapter.

There are increasing number of studies claiming that neurogenesis is possible in the adult central nervous system[115,116] some even raising hope for brain 'rejuvenation'.[117] Most of these claims are based on the evidence that these neurones are able to replicate their DNA. The discovery of the cell cycle and a more thorough understanding of the regulatory mechanisms impose a strict distinction between DNA replication and cell division. DNA replication indeed precedes mitosis but mitosis does not always follow DNA replication. This means that BrdU incorporation in neurones does not necessarily indicate neurogenesis. As much as I am optimistic about possible treatments for dementia, in the short term I believe in preventing cell loss rather than hope that neurones can be replaced.

References

1. Malumbres M, Barbacid M. To cycle or not to cycle: A critical decision in cancer. Nature reviews. Cancer 2001; 1(3):222-231.
2. Uchiyama Y. Apoptosis: The history and trends of its studies. Archives of Histology and Cytology 1995; 58(2):127-137.
3. Wilson MR. Apoptotic signal transduction: Emerging pathways. Biochemistry and Cell Biology 1998; 76(4):573-582.
4. Blaschke AJ, Staley K, Chun J. Widespread programmed cell death in proliferative and postmitotic regions of the fetal cerebral cortex. Development 1996; 122(4):1165-1174.
5. Nguyen MD, Mushynski WE, Julien JP. Cycling at the interface between neurodevelopment and neurodegeneration. Cell Death and Differentiation 2002; 9(12):1294-1306.
6. Sommer L, Rao M. Neural stem cells and regulation of cell number. Progress in Neurobiology 2002; 66(1):1-18.
7. Peunova N, Scheinker V, Cline H et al. Nitric oxide is an essential negative regulator of cell proliferation in Xenopus brain. J Neurosci 2001; 21(22):8809-8818.
8. Groszer M, Erickson R, Scripture-Adams DD et al. Negative regulation of neural stem/progenitor cell proliferation by the Pten tumor suppressor gene in vivo. Science 2001; 294(5549):2186-2189.
9. Quinn LM, Herr A, McGarry TJ et al. The Drosophila Geminin homolog: Roles for Geminin in limiting DNA replication, in anaphase and in neurogenesis. Genes Dev 2001; 15(20):2741-2754.
10. Mitchell BD, Gibbons B, Allen LR et al. Aberrant apoptosis in the neurological mutant flathead is associated with defective cytokinesis of neural progenitor cells. Brain Res Dev Brain Res 2001; 130(1):53-63.
11. Olson JM, Asakura A, Snider L et al. NeuroD2 is necessary for development and survival of central nervous system neurons. Dev Biol 2001; 234(1):174-187.
12. Gomes WA, Kessler JA. Msx-2 and p21 mediate the pro apoptotic but not the anti-proliferative effects of BMP4 on cultured sympathetic neuroblasts. Dev Biol 2001; 237(1):212-221.
13. Lossi L, Coli A, Giannessi E et al. Cell proliferation and apoptosis during histogenesis of the guinea pig and rabbit cerebellar cortex. Italian journal of anatomy and embryology. Archivio Italiano di Anatomia ed Embriologia 2002; 107(2):117-125.
14. Erhardt JA, Pittman RN. p21WAF1 induces permanent growth arrest and enhances differentiation, but does not alter apoptosis in PC12 cells. Oncogene 1998; 16(4):443-451.

15. Park DS, Levine B, Ferrari G et al. Cyclin dependent kinase inhibitors and dominant negative cyclin dependent kinase 4 and 6 promote survival of NGF-deprived sympathetic neurons. J Neurosci 1997; 17(23):8975-8983.

16. Mani S, Shen Y, Schaefer J et al. Failure to express GAP-43 during neurogenesis affects cell cycle regulation and differentiation of neural precursors and stimulates apoptosis of neurons. Mol Cell Neurosci 2001; 17(1):54-66.

17. Gonzalez Garcia M, Garcia I, Ding L et al. Bcl-x is expressed in embryonic and postnatal neural tissues and functions to prevent neuronal cell death. Proceedings of the National Academy of Sciences of the United States of America 1995; 92(10):4304-4308.

18. Castren E, Ohga Y, Berzaghi MP et al. Bcl-2 messenger RNA is localized in neurons of the developing and adult rat brain. Neuroscience 1994; 61(1):165-177.

19. Chen DF, Schneider GE, Martinou JC et al. Bcl-2 promotes regeneration of severed axons in mammalian CNS. Nature 1997; 385(6615):434-439.

20. Daadi MM, Saporta S, Willing AE et al. In vitro induction and in vivo expression of bcl-2 in the hNT neurons. Brain Research Bulletin 2001; 56(2):147-152.

21. Hilton M, Middleton G, Davies AM. Bcl-2 influences axonal growth rate in embryonic sensory neurons. Current Biology CB 1997; 7(10):798-800.

22. Vairo G, Soos TJ, Upton TM et al. Bcl-2 retards cell cycle entry through p27(Kip1), pRB relative p130, and altered E2F regulation. Molecular and Cellular Biology 2000; 20(13):4745-4753.

23. Cheng A, Chan SL, Milhavet O et al. p38 MAP kinase mediates nitric oxide-induced apoptosis of neural progenitor cells. J Biol Chem 2001; 276(46):43320-43327.

24. Lee Y, Chong MJ, McKinnon PJ. Ataxia telangiectasia mutated-dependent apoptosis after genotoxic stress in the developing nervous system is determined by cellular differentiation status. J Neurosci 2001; 21(17):6687-6693.

25. Eves EM, Boise LH, Thompson CB et al. Apoptosis induced by differentiation or serum deprivation in an immortalized central nervous system neuronal cell line. Journal of Neurochemistry 1996; 67(5):1908-1920.

26. Noble M, Dietrich J. Intersections between neurobiology and oncology: Tumor origin, treatment and repair of treatment-associated damage. Trends Neurosci 2002; 25(2):103-107.

27. Roman DD, Sperduto PW. Neuropsychological effects of cranial radiation: Current knowledge and future directions. Int J Radiat Oncol Biol Phys 1995; 31(4):983-998.

28. Radcliffe J, Bunin GR, Sutton LN et al. Cognitive deficits in long-term survivors of childhood medulloblastoma and other noncortical tumors: Age-dependent effects of whole brain radiation. Int J Dev Neurosci 1994; 12(4):327-334.

29. Schagen SB, van Dam FS, Muller MJ et al. Cognitive deficits after postoperative adjuvant chemotherapy for breast carcinoma. Cancer 1999; 85(3):640-650.

30. van Dam FS, Schagen SB, Muller MJ et al. Impairment of cognitive function in women receiving adjuvant treatment for high-risk breast cancer: High-dose versus standard-dose chemotherapy. J Natl Cancer Inst 1998; 90(3):210-218.

31. Kubota Y, Takahashi S, Sun XZ et al. Radiation-induced tissue abnormalities in fetal brain are related to apoptosis immediately after irradiation. Int J Radiat Biol 2000; 76(5):649-659.

32. Peissner W, Kocher M, Treuer H et al. Ionizing radiation-induced apoptosis of proliferating stem cells in the dentate gyrus of the adult rat hippocampus. Brain Res Mol Brain Res 1999; 71(1):61-68.

33. Bellinzona M, Gobbel GT, Shinohara C et al. Apoptosis is induced in the subependyma of young adult rats by ionizing irradiation. Neurosci Lett 1996; 208(3):163-166.

34. Uberti D, Piccioni L, Cadei M et al. p53 is dispensable for apoptosis but controls neurogenesis of mouse dentate gyrus cells following gamma-irradiation. Brain Res Mol Brain Res 2001; 93(1):81-89.

35. Gobbel GT, Bellinzona M, Vogt AR et al. Response of postmitotic neurons to X-irradiation: Implications for the role of DNA damage in neuronal apoptosis. J Neurosci 1998; 18(1):147-155.

36. Tikka T, Usenius T, Tenhunen M et al. Tetracycline derivatives and ceftriaxone, a cephalosporin antibiotic, protect neurons against apoptosis induced by ionizing radiation. J Neurochem 2001; 78(6):1409-1414.

37. Bendeck MP, Conte M, Zhang M et al. Doxycycline modulates smooth muscle cell growth, migration, and matrix remodeling after arterial injury. Am J Pathol 2002; 160(3):1089-1095.

38. Park DS, Morris EJ, Padmanabhan J et al. Cyclin-dependent kinases participate in death of neurons evoked by DNA-damaging agents. J Cell Biol 1998; 143(2):457-467.

39. Fischer SJ, McDonald ES, Gross L et al. Alterations in cell cycle regulation underlie cisplatin induced apoptosis of dorsal root ganglion neurons in vivo. Neurobiol Dis 2001; 8(6):1027-1035.

40. Lee HW, Koo H, Choi KG et al. The effects of peripheral leukocytes on the hippocampal neuronal changes in transient global ischemia and unilateral cerebral hemispheric infarction. J Korean Med Sci 1999; 14(3):304-314.

41. Sebille A, Bondoux-Jahan M. Motor function recovery after axotomy: Enhancement by cyclophosphamide and spermine in rat. Exp Neurol 1980; 70(3):507-515.

42. Husseman JW, Nochlin D, Vincent I. Mitotic activation: A convergent mechanism for a cohort of neurodegenerative diseases. Neurobiology of Aging 2000; 21(6):815-828.

43. Arendt T, Rodel L, Gartner U et al. Expression of the cyclin-dependent kinase inhibitor p16 in Alzheimer's disease. Neuroreport 1996; 7(18):3047-3049.

44. Nagy Zs, Esiri MM, Cato AM et al. Cell cycle markers in the hippocampus in Alzheimer's disease. Acta Neuropathol (Berl) 1997; 94(1):6-15.

45. Nagy Zs, Esiri MM, Smith AD. Expression of cell division markers in the hippocampus in Alzheimer's disease and other neurodegenerative conditions. Acta Neuropathol (Berl) 1997; 93(3):294-300.

46. Vincent I, Rosado M, Davies P. Mitotic mechanisms in Alzheimer's disease? J Cell Biol 1996; 132(3):413-425.

47. Smith TW, Lippa CF. Ki-67 immunoreactivity in Alzheimer's disease and other neurodegenerative disorders. J Neuropathol Exp Neurol 1995; 54(3):297-303.

48. Nagy Zs, Esiri MM. Neuronal cyclin expression in the hippocampus in temporal lobe epilepsy. Exp Neurol 1998; 150(2):240-247.

49. Vincent I, Jicha G, Rosado M et al. Aberrant expression of mitotic cdc2/cyclin B1 kinase in degenerating neurons of Alzheimer's disease brain. J Neurosci 1997; 17(10):3588-3598.

50. Nagy Zs, Esiri MM. Apoptosis-related protein expression in the hippocampus in Alzheimer's disease. Neurobiol Aging 1997; 18(6):565-571.

51. Busser J, Geldmacher DS, Herrup K. Ectopic cell cycle proteins predict the sites of neuronal cell death in Alzheimer's disease brain. Journal of Neuroscience the Official Journal of the Society for Neuroscience 1998; 18(8):2801-2807.

52. Nagy Zs, Esiri MM, Smith AD. The cell division cycle and the pathophysiology of Alzheimer's disease. Neuroscience 1998; 87(4):731-739.

53. Hamdane M, Smet C, Sambo AV et al. Pin1: A therapeutic target in Alzheimer neurodegeneration. Journal of Molecular Neuroscience MN 2002; 19(3):275-287.

54. Hoozemans JJ, Bruckner MK, Rozemuller AJ et al. Cyclin D1 and cyclin E are colocalized with cyclo-oxygenase 2 (COX-2) in pyramidal neurons in Alzheimer disease temporal cortex. Journal of Neuropathology and Experimental Neurology 2002; 61(8):678-688.

55. Yang Y, Geldmacher DS, Herrup K. DNA replication precedes neuronal cell death in Alzheimer's disease. Journal of Neuroscience the Official Journal of the Society for Neuroscience 2001; 21(8):2661-2668.

56. Ding XL, Husseman J, Tomashevski A et al. The cell cycle Cdc25A tyrosine phosphatase is activated in degenerating postmitotic neurons in Alzheimer's disease. American Journal of Pathology 2000; 157(6):1983-1990.

57. Zhu X, Rottkamp CA, Raina AK et al. Neuronal CDK7 in hippocampus is related to aging and Alzheimer disease. Neurobiology of Aging 2000; 21(6):807-813.

58. Tsujioka Y, Takahashi M, Tsuboi Y et al. Localization and expression of cdc2 and cdk4 in Alzheimer brain tissue. Dementia and Geriatric Cognitive Disorders 1999; 10(3):192-198.

59. Giovanni A, Wirtz Brugger F, Keramaris E et al. Involvement of cell cycle elements, cyclin-dependent kinases, pRb, and E2F x DP, in B-amyloid-induced neuronal death. Journal of Biological Chemistry 1999; 274(27):19011-19016.

60. Arendt T, Holzer M, Gartner U. Neuronal expression of cycline dependent kinase inhibitors of the INK4 family in Alzheimer's disease. Journal of Neural Transmission Vienna Austria 1996 1998; 105(8-9):949-960.

61. McShea A, Harris PL, Webster KR et al. Abnormal expression of the cell cycle regulators P16 and CDK4 in Alzheimer's disease. American Journal of Pathology 1997; 150(6):1933-1939.

62. Ferrer I, Blanco R, Carmona M et al. Phosphorylated c-MYC expression in Alzheimer disease, Pick's disease, progressive supranuclear palsy and corticobasal degeneration. Neuropathology and Applied Neurobiology 2001; 27(5):343-351.

63. Raina AK, Pardo P, Rottkamp CA et al. Neurons in Alzheimer disease emerge from senescence. Mech Ageing Dev 2001; 123(1):3-9.

64. Arendt T. Alzheimer's disease as a loss of differentiation control in a subset of neurons that retain immature features in the adult brain. Neurobiology of Aging 2000; 21(6):783-796.

65. Arendt T. Alzheimer's disease as a disorder of mechanisms underlying structural brain self-organization. Neuroscience 2001; 102(4):723-765.

66. Arendt T. Disturbance of neuronal plasticity is a critical pathogenetic event in Alzheimer's disease. International Journal of Developmental Neuroscience the Official Journal of the International Society for Developmental Neuroscience 2001; 19(3):231-245.

67. Nagy Zs. Cell cycle regulatory failure in neurones: Causes and consequences. Neurobiology of Aging 2000; 21(6):761-769.

68. Arendt T, Schindler C, Bruckner MK et al. Plastic neuronal remodeling is impaired in patients with Alzheimer's disease carrying apolipoprotein epsilon 4 allele. Journal of Neuroscience the Official Journal of the Society for Neuroscience 1997; 17(2):516-529.

69. Lustig RH. Sex hormone modulation of neural development in vitro. Hormones and Behavior 1994; 28(4):383-395.

70. Mussa GC, Mussa F, Bretto R et al. Influence of thyroid in nervous system growth. Minerva Pediatrica 2001; 53(4):325-353.

71. Jordan Sciutto KL, Malaiyandi LM, Bowser R. Altered distribution of cell cycle transcriptional regulators during Alzheimer disease. Journal of Neuropathology and Experimental Neurology 2002; 61(4):358-367.

72. Suzuki T, Oishi M, Marshak DR et al. Cell cycle-dependent regulation of the phosphorylation and metabolism of the Alzheimer amyloid precursor protein. EMBO Journal 1994; 13(5):1114-1122.

73. Preuss U, Doring F, Illenberger S et al. Cell cycle-dependent phosphorylation and microtubule binding of tau protein stably transfected into Chinese hamster ovary cells. Molecular Biology of the Cell 1995; 6(10):1397-1410.

74. Pope WB, Lambert MP, Leypold B et al. Microtubule-associated protein tau is hyperphosphorylated during mitosis in the human neuroblastoma cell line SH-SY5Y. Experimental Neurology 1994; 126(2):185-194.

75. Bancher C, Lassmann H, Breitschopf H et al. Mechanisms of cell death in Alzheimer's disease. Journal of Neural Transmission. Supplementum 1997; 50 141-152.

76. Stadelmann C, Deckwerth TL, Srinivasan A et al. Activation of caspase-3 in single neurons and autophagic granules of granulovacuolar degeneration in Alzheimer's disease. Evidence for apoptotic cell death. American Journal of Pathology 1999; 155(5):1459-1466.

77. Jellinger KA, Stadelmann CH. The enigma of cell death in neurodegenerative disorders. Journal of neural transmission. Supplementum 2000; (60):21-36.

78. Kosel S, Egensperger R, von Eitzen U et al. On the question of apoptosis in the parkinsonian substantia nigra. Acta Neuropathologica 1997; 93(2):105-108.

79. Zhu X, Rottkamp CA, Boux H et al. Activation of p38 kinase links tau phosphorylation, oxidative stress, and cell cycle-related events in Alzheimer disease. J Neuropathol Exp Neurol 2000; 59(10):880-888.

80. Paganelli AR, Ocana OH, Prat MI et al. The Alzheimer-related gene presenilin-1 facilitates sonic hedgehog expression in Xenopus primary neurogenesis. Mech Dev 2001; 107(1-2):119-131.

81. Janicki SM, Stabler SM, Monteiro MJ. Familial Alzheimer's disease presenilin-1 mutants potentiate cell cycle arrest. Neurobiology of Aging 2000; 21(6):829-836.

82. Neve RL, McPhie DL, Chen Y. Alzheimer's disease: Dysfunction of a signalling pathway mediated by the amyloid precursor protein? Biochem Soc Symp 2001; 67 37-50.

83. Copani A, Condorelli F, Caruso A et al. Mitotic signaling by beta-amyloid causes neuronal death. Faseb J 1999; 13(15):2225-2234.

84. Chen Y, McPhie DL, Hirschberg J et al. The amyloid precursor protein-binding protein APP-BP1 drives the cell cycle through the S-M checkpoint and causes apoptosis in neurons. J Biol Chem 2000; 275(12):8929-8935.

85. Giovanni A, Keramaris E, Morris EJ et al. E2F1 mediates death of B-amyloid-treated cortical neurons in a manner independent of p53 and dependent on Bax and caspase 3. J Biol Chem 2000; 275(16):11553-11560.
86. Wu Q, Combs C, Cannady SB et al. Beta-amyloid activated microglia induce cell cycling and cell death in cultured cortical neurons. Neurobiol Aging 2000; 21(6):797-806.
87. Katchanov J, Harms C, Gertz K et al. Mild cerebral ischemia induces loss of cyclin-dependent kinase inhibitors and activation of cell cycle machinery before delayed neuronal cell death. J Neurosci 2001; 21(14):5045-5053.
88. Guegan C, Levy V, David JP et al. c-Jun and cyclin D1 proteins as mediators of neuronal death after a focal ischaemic insult. Neuroreport 1997; 8(4):1003-1007.
89. Small DL, Monette R, Fournier MC et al. Characterization of cyclin D1 expression in a rat global model of cerebral ischemia. Brain Res 2001; 900(1):26-37.
90. Li Y, Chopp M, Powers C et al. Apoptosis and protein expression after focal cerebral ischemia in rat. Brain Res 1997; 765(2):301-312.
91. Li Y, Chopp M, Powers C et al. Immunoreactivity of cyclin D1/cdk4 in neurons and oligodendrocytes after focal cerebral ischemia in rat. J Cereb Blood Flow Metab 1997; 17(8):846-856.
92. Li Y, Chopp M, Powers C. Granule cell apoptosis and protein expression in hippocampal dentate gyrus after forebrain ischemia in the rat. J Neurol Sci 1997; 150(2):93-102.
93. Hayashi T, Sakai K, Sasaki C et al. Phosphorylation of retinoblastoma protein in rat brain after transient middle cerebral artery occlusion. Neuropathol Appl Neurobiol 2000; 26(4):390-397.
94. Bossenmeyer-Pourie C, Koziel V, Daval JL. CPP32/CASPASE-3-like proteases in hypoxia-induced apoptosis in developing brain neurons. Brain Res Mol Brain Res 1999; 71(2):225-237.
95. Bossenmeyer-Pourie C, Chihab R, Schroeder H et al. Transient hypoxia may lead to neuronal proliferation in the developing mammalian brain: From apoptosis to cell cycle completion. Neuroscience 1999; 91(1):221-231.
96. Timsit S, Rivera S, Ouaghi P et al. Increased cyclin D1 in vulnerable neurons in the hippocampus after ischaemia and epilepsy: A modulator of in vivo programmed cell death? Eur J Neurosci 1999; 11(1):263-278.
97. Tomasevic G, Kamme F, Stubberod P et al. The tumor suppressor p53 and its response gene p21WAF1/Cip1 are not markers of neuronal death following transient global cerebral ischemia. Neuroscience 1999; 90(3):781-792.
98. Esiri MM, Nagy Zs, Smith MZ et al. Cerebrovascular disease and threshold for dementia in the early stages of Alzheimer's disease. Lancet 1999; 354(9182):919-920.
99. Nagy Zs, Esiri MM, Jobst KA et al. The effects of additional pathology on the cognitive deficit in Alzheimer disease. Journal of Neuropathology and Experimental Neurology 1997; 56(2):165-170.
100. Smith MZ, Nagy Zs, Esiri MM. Cell cycle-related protein expression in vascular dementia and Alzheimer's disease. Neuroscience Letters 1999; 271(1):45-48.
101. Smith MZ, Nagy Zs, Barnetson L et al. Coexisting pathologies in the brain: Influence of vascular disease and Parkinson's disease on Alzheimer's pathology in the hippocampus. Acta Neuropathologica 2000; 100(1):87-94.
102. Houser CR. Morphological changes in the dentate gyrus in human temporal lobe epilepsy. Epilepsy Res Suppl 1992; 7 223-234.
103. Pollard H, Cantagrel S, Charriaut-Marlangue C et al. Apoptosis associated DNA fragmentation in epileptic brain damage. Neuroreport 1994; 5(9):1053-1055.
104. Dragunow M, Faull RL, Lawlor P et al. In situ evidence for DNA fragmentation in Huntington's disease striatum and Alzheimer's disease temporal lobes. Neuroreport 1995; 6(7):1053-1057.
105. Bengzon J, Kokaia Z, Elmer E et al. Apoptosis and proliferation of dentate gyrus neurons after single and intermittent limbic seizures. Proc Natl Acad Sci USA 1997; 94(19):10432-10437.
106. Zhang LX, Smith MA, Li XL et al. Apoptosis of hippocampal neurons after amygdala kindled seizures. Brain Res Mol Brain Res 1998; 55(2):198-208.
107. Ino H, Chiba T. Cyclin-dependent kinase 4 and cyclin D1 are required for excitotoxin-induced neuronal cell death in vivo. J Neurosci 2001; 21(16):6086-6094.
108. Chen G, Zeng W, Yuan P et al. The mood-stabilizing agents lithium and valproate robustly increase the levels of the neuroprotective protein bcl-2 in the CNS. J Neurochem 1999; 72(2):879-882.

109. Madsen TM, Treschow A, Bengzon J et al. Increased neurogenesis in a model of electroconvulsive therapy. Biol Psychiatry 2000; 47(12):1043-1049.

110. Reid IC, Stewart CA. Seizures, memory and synaptic plasticity. Seizure 1997; 6(5):351-359.

111. Nagy Zs, Combrinck M, Budge M et al. Cell cycle kinesis in lymphocytes in the diagnosis of Alzheimer's disease. Neuroscience Letters 2002; 317(2):81-84.

112. Stieler JT, Lederer C, Bruckner MK et al. Impairment of mitogenic activation of peripheral blood lymphocytes in Alzheimer's disease. Neuroreport 2001; 12(18):3969-3972.

113. Uberti D, Carsana T, Bernardi E et al. Selective impairment of p53-mediated cell death in fibroblasts from sporadic Alzheimer's disease patients. Journal of Cell Science 2002; 115(Pt 15):3131-3138.

114. Dhavan R, Tsai LH. A decade of CDK5. Nature reviews. Molecular Cell Biology 2001; 2(10):749-759.

115. Lasley EN. Neurobiology. Death leads to brain neuron birth. Science 2000; 288(5474):2111-2112.

116. Magavi SS, Leavitt BR, Macklis JD. Induction of neurogenesis in the neocortex of adult mice. Nature 2000; 405(6789):951-955.

117. Nottebohm F. Neuronal replacement in adult brain. Brain Research Bulletin 2002; 57(6):737-749.

CHAPTER 5

Cell Cycle and Chromosome Segregation Defects in Alzheimer's Disease

Huntington Potter

Abstract

Despite a common set of hallmark neuropathological lesions and clinical symptoms, Alzheimer's disease has an apparently complex etiology. The disease can be caused by autosomal dominant mutations in at least three genes (encoding the amyloid precursor protein (APP) and the two presenilins). In addition, it can be influenced by certain allelic variants of at least three "risk factor" genes (apolipoprotein E, antichymotrypsin, and interleukin-1), or may arise "sporadically" with no evident genetic components. In the end, as many as 30-40% of individuals over the age of 85 may have some symptoms of Alzheimer's—underscoring the fact that age itself is the strongest risk factor for the disease.

It has been known for almost twenty years that individuals with trisomy 21 (Down syndrome) exhibit Alzheimer neuropathology by the time they are 30-40 years old. Somewhat later, they also develop dementia, and eventually die of Alzheimer's disease. Because the gene for amyloid precursor protein (APP) resides on chromosome 21, its consequent overexpression in trisomy 21 cells presumably contributes to the development of Alzheimer's disease in Down syndrome individuals.

The connection between Down syndrome and Alzheimer's disease and the application of Occam's Razor led me to hypothesize that many cases of classical Alzheimer's disease—both of the genetic and late-onset, sporadic forms—might similarly be caused by chromosome mis-segregation leading to a small number of trisomy 21 cells developing during the life of the affected individual.

In this chapter, I will consider evidence from several laboratories that defects in mitosis, and particularly in chromosome segregation, may be a part of the Alzheimer disease process. In particular, mutations in the presenilin genes that cause Alzheimer's disease also cause chromosome instability. By generating a mosaic population of trisomy 21 and other aneuploid cells, such a mitotic defect could lead to Alzheimer pathology and dementia by inducing inflammation, apoptosis, and/or altered processing of the APP protein into the neurotoxic amyloid β-protein—all characteristic features of the disease. The possibility that many cases of Alzheimer's disease are mosaic for trisomy 21 suggests novel approaches to diagnosis and therapy.

Introduction

Alzheimer's disease arises when neurons in certain regions of the brain, particularly those involved in memory and cognition, are damaged and ultimately killed. A key step in this

Cell Cycle Mechanisms and Neuronal Cell Death, edited by Agata Copani and Ferdinando Nicoletti. ©2005 Eurekah.com and Kluwer Academic / Plenum Publishers.

process is the polymerization of the Aβ peptide into neurotoxic protein filaments. Aggregates of these filaments accumulate in the brain as the characteristic neuropathological lesions termed "amyloid" and are thought to be an essential contributor to neuronal cell death in Alzheimer's disease.[1-5] This hypothesis has been supported by many in vitro and in vivo experiments[6-12] and has been recently strengthened greatly by our demonstration that amyloid formation catalyzed by the action of two amyloid-associated proteins, apolipoprotein E (apoE) and antichymotrypsin (ACT), on Aβ is required for neuronal disfunction and cognitive impairment in a mouse model of Alzheimer's disease.[12] Together with the genetic evidence implicating the amyloid precursor protein (APP), apoE4, and ACT-A in the disease (for review, see reference 5), these results strongly point to either the process or product of amyloid formation as the key to Alzheimer's disease. The unanswered questions are how does amyloid formation arise in the majority of non-inherited Alzheimer's disease and what initiates neuronal cell death.

An important clue to the mechanism of Alzheimer's disease was the discovery that Down syndrome patients who live beyond the age of 30 or 40 develop brain neuropathology indistinguishable from that observed in classical Alzheimer's disease.[13-15] Down syndrome is caused by the presence of three copies of chromosome 21, instead of the usual two, in every cell of the body from the moment of conception. The implication of this finding is that trisomy for chromosome 21 not only causes Down syndrome, but is also sufficient to cause Alzheimer's disease later in life.[16]

One possible explanation for the link between Down syndrome and Alzheimer's disease is that, in both disorders, a gene on chromosome 21 is over-expressed—due either to the 50% increased dosage of chromosome 21 genes in Down syndrome, or to potential somatic or inherited mutation or gene duplication in Alzheimer's disease.[17,18] Indeed, the Aβ peptide, which the major component of the pathological amyloid deposits found in Alzheimer's disease brain, is encoded by a gene (amyloid precursor protein; APP) that resides on chromosome 21.[19-22] APP is actually over-expressed in Down syndrome individuals somewhat more than the 50% expected from gene dosage alone.[23-25] However, no known cases of Alzheimer's disease have resulted from a simple duplication of the APP gene or over expression of APP. It therefore seems to me more likely that trisomy 21 results in multiple abnormalities in gene expression that together result in Alzheimer's disease decades later.

One weekend, some years ago, I was deep in the New Hampshire woods attending a retreat for the MD-PhD students and faculty of The Harvard Medical School. Taking advantage of a relaxed atmosphere, I tried to concentrate on the curious relationship between Alzheimer's disease and Down syndrome. It occurred to me that if complete "trisomy 21" could lead to early Alzheimer's disease in Down syndrome, perhaps the slow development of some trisomy 21 cells over a lifetime could cause the Alzheimer's disease that affects elderly individuals.[26] The more I thought about it the more it became clear that this model explained many, seemingly unrelated facts about Alzheimer's disease. It could also account for both the inherited and more common, non-familial form of the disorder, depending upon whether the defect in chromosome segregation that led to trisomy 21 mosaicism was the result of a genetic mutation or some environmental insult.

In order to be useful, a scientific model must be able to make testable predictions. The trisomy 21 model for Alzheimer's disease makes at least two major ones. The first is that Down syndrome and Alzheimer patients should share clinical features besides dementia that might be used as a diagnostic test for Alzheimer's disease.[26] One such potential test is based on the finding that Down syndrome and Alzheimer's disease individuals have cholinergic deficits that, for instance make them hypersensitive to the pupil dilating effect of cholinergic antagonists.[26,27] We and others are still in the process of testing this prediction. Increasingly promising results suggest that even individuals that are seemingly cognitively normal can dilate abnormally in response to the cholinergic antagonist tropicamide and can show cholinergic pathology and neuronal cell death not only in the cortex and hippocampus, but also in the Edinger-Westphal nucleus that controls the pupil.[28-30] Such individuals are likely to be in the preclinical stage of Alzheimer's disease, which can last 10-30 years or more, and may be detected by their hypersensitive pupil dilation in response to tropicamide.

The second, and most important prediction of the trisomy 21 model of Alzheimer's disease is that Alzheimer patients should have developed a small number of trisomy 21 cells in their somatic tissues.[26] Such trisomy 21 mosaicism would be produced over time by unequal chromosome segregation during mitosis. Ultimately, trisomy 21 cells in the mosaic individual could lead to Alzheimer's disease through the same (as yet unknown, and presumably multi-step) mechanism by which Down syndrome patients acquire the disease, but at a later age due to the modulating effect of the majority population of normal diploid cells.

Based on this suggestion that Alzheimer's disease may be a mosaic form of Down syndrome arising from defects in mitosis, certain further predictions could be made:[26]

1. There should be alterations in the mitotic spindle apparatus or in mitosis-related proteins in Alzheimer's disease cells that could lead to chromosome mis-segregation and trisomy 21 mosaicism, and
2. Mutations that cause Alzheimer's disease should occur in genes coding for proteins directly or indirectly involved in mitosis and chromosome segregation.

Each of these predictions can be tested, either by re-examination of the literature for findings supportive of the theory, or by new experimentation. In this chapter, we will first examine the published and unpublished evidence in favor of chromosome instability/ mis-segregation and trisomy 21 mosaicism as a cause of Alzheimer's disease, then present evidence linking the Alzheimer presenilin mutations to mitosis and chromosome mis-segregation, and finally consider the resulting implications for Alzheimer diagnosis and therapy.

Trisomy 21 Mosaicism in Alzheimer's Disease

An early hint that chromosome mis-segregation and trisomy 21 mosaicism might be related to Alzheimer's disease was provided by epidemiological studies showing that women in some Alzheimer families in which the disease is inherited as an autosomal dominant mutation give birth to a significantly higher-than-normal number of Down syndrome children.[31-33] Other researchers have failed to confirm the increased incidence of Down syndrome in families with inherited Alzheimer's disease, but they report that the number of relatives they analyzed was too few for the lack of Down syndrome to be statistically significant.[34-36] Although the data in the largest studies were very suggestive, more work needs to be carried out to confirm the connection between a high frequency of Down syndrome and inherited Alzheimer's disease in the same family.

More recently, a strong association between Down syndrome and sporadic Alzheimer's disease was found. In a retrospective study, Schupf and her colleagues showed that young mothers of Down syndrome children have a five-fold greater risk of developing Alzheimer's disease later in life, when compared to either older Down syndrome mothers, or the general population, as though they suffered a novel form of "accelerated aging".[37] We interpret this result, instead, as indicating that the young Down syndrome mothers were most likely mosaic for trisomy 21 (as if they had a predisposition to chromosome mis-segregation) and that this mosaicism was reflected in their trisomy 21 offspring and their own increased risk for Alzheimer's disease.

Finally, trisomy 21 mosaicism was specifically found in two women who, though not mentally retarded and not characterized as having Down syndrome, developed Alzheimer-like dementia by age 40.[38-40] One of the two women also had a Down syndrome child. An unusual family with an inherited aberrant chromosome 22-derived mini-chromosome was studied by Percy et al and found, in addition, to have a high frequency of Alzheimer's disease.[41] The two living Alzheimer-affected members of the family carried the marker chromosome and one was also found to be mosaic for trisomy 21. The two patients reported by Schapiro[38] and Rowe[39] and their colleagues, and possibly the mosaic individual reported by Percy[41] and her colleagues, demonstrate that it is not necessary for every cell of an individual to be trisomy 21 for the aberrant effects of this chromosome imbalance to result in early Alzheimer dementia. The often-later onset dementia of classic Alzheimer's disease—both genetic and sporadic—could thus result from an even smaller percentage of trisomy 21 cells that may go undetected.

It would be clearly advantageous to determine directly whether Alzheimer's disease patients are mosaic for trisomy 21. However, this apparently simple experiment is not straightforward because techniques must be used that can accurately determine low levels of mosaicism. Cytogenetic analysis of Alzheimer's disease patients has been carried out in a number of laboratories, with mixed results. Some investigators have reported small increases in aneuploidy or other abnormalities as measured by direct karyotyping, while others failed to confirm the finding, especially if they analyzed few or small families.[42-50] Interestingly, the cytogenetic abnormality termed premature centromere division (PCD), which is a potential cause of improper chromosome segregation in vitro and in vivo, was found to be positively correlated with age and to be increased in women with familial Alzheimer's disease (3.6% vs. 0.6% in age-matched controls), particularly affecting the X chromosome.[47,50,51] Furthermore, trisomy 21, 18, and X occurred in the lymphocytes and fibroblasts of a woman prone to PCD, who also had three trisomy 21 conceptuses.[52]

In contrast to metaphase karyotyping, fluorescence in situ hybridization (FISH) allows the number of copies of a particular chromosome to be determined regardless of the stage of the cell cycle (mitotic or interphase).[53,54] The ability to examine interphase nuclei also greatly increases the number of cells that can be analyzed and, in principle, should allow easier detection of very low levels of aneuploidy. We therefore used FISH to determine the extent of aneuploidy, particularly chromosome 21 trisomy, in fibroblasts from Alzheimer and normal individuals.[1,55,56] In analyzing thousands of cells from 27 Alzheimer and 13 control individuals, we found that fibroblasts from Alzheimer's disease patients exhibited more than twice the frequency of trisomy 21 as did cells from age-matched, normal individuals (Fig. 1). The average percentage of trisomy 21 cells was 5.5% in Alzheimer's disease cultures and 2.5% in cultures from unaffected individuals. Using the same procedure, the number of trisomies in a Down syndrome culture was determined to be 98%. The greater frequency of trisomy 21 cells in Alzheimer's disease patients compared to controls was significant (p=0.007) and was not related to the age of the affected individuals. A small parallel study of chromosome 18 showed similar aneuploidy, indicating that the chromosome mis-segregation likely affected all chromosomes.

When the chromosome study was initiated, we decided to examine all types of Alzheimer's disease with a particular focus on individuals with the early onset, familial form of the disease. The majority of these cases were, in the course of the investigation, shown to carry mutations in one of the two presenilin genes.[57-62] We found that chromosome mis-segregation had occurred in all Alzheimer's disease individuals, including those with sporadic AD and those who are now know to carry a familial Alzheimer's disease (FAD)-causing mutation in either presenilin 1 or presenilin 2. Chromosome mis-segregation in sporadic AD patients has been elegantly confirmed in blood lymphocytes by Migliore et al[63] and in the brains of sporadic AD patients by Yang and colleagues.[64]

Defects in Mitosis in Alzheimer's Disease

During the course of the studies described above, several lines of investigation provided independent evidence that defects in mitosis and/or in mitosis-specific proteins may be present in Alzheimer's disease patients—supporting another prediction of the trisomy 21 mosaicism model. Such defects could be expected to lead to chromosome mis-segregation, and thus could

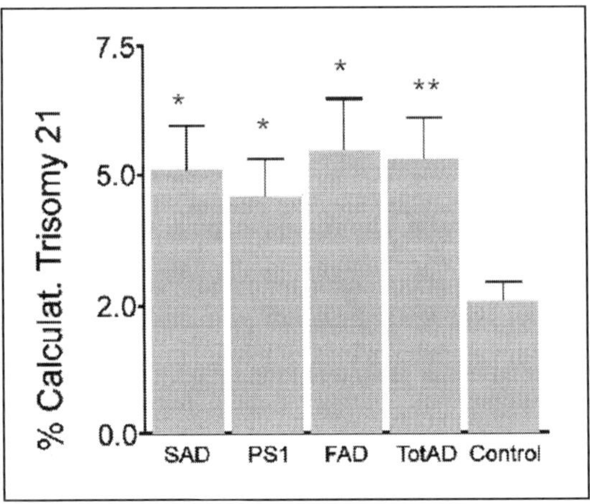

Figure 1. AD cells, including those with FAD mutations in PS-1 or 2 show increased trisomy 21 aneuploidy. Reprinted with permission from Neurobiology of Disease 1999; 6:167-179 ©1999 Elsevier.[56]

result in the trisomy 21 mosaicism and other aneuploidy observed in Alzheimer' disease cells and individuals, and, as will be discussed, may also lead to other features of the disease such as apoptosis and changes in amyloid precursor protein (APP) processing.

The most direct evidence for mitotic defects in Alzheimer's disease has been provided by experiments in which the mitotic spindles in dividing cells from Alzheimer patients were observed to be abnormal. For example, we briefly treated lymphocyte cell lines from presenilin mutant Alzheimer's disease and control patients with the microtubule-disrupting agent colchicine, and analyzing their karyotypes 40 hours later. This treatment causes many of the cells to exhibit separated chromatids in metaphase spreads. That is, in these karyotypes, the individual chromatids lay parallel to each other, separated by a clear gap, rather than being connected at their centromeres, as are chromosomes from untreated cultures. This metaphase chromosome pattern is similar to the spontaneous premature centromere division (PCD) that made the patients described above prone to chromosome mis-segregation and Alzheimer dementia. Comparison of colchicine-treated cells showed that those from Alzheimer's disease patients exhibited this karyotypic abnormality significantly more frequently than cells from normal individuals of similar age[1] (Geller, Benjamin and Potter, in preparation). A similar Alzheimer-specific increase in PCD induced by microtubule-disrupting agents was reported by Migliori and colleagues.[50]

A related defect in mitosis (increased frequency of chromosomes displaced from the mitotic spindle after colchicine treatment) was found by Ford in cells derived from mothers of Down syndrome children.[65] In addition, the rate of hyperdiploidy of all chromosomes was increased in cells from mothers with a Down syndrome child.[66] These data indicate that such mothers may be generally prone to chromosome mis-segregation, which can lead to either gonadal trisomy 21 mosaicism or to meiotic non-disjunction, or both, and thus resulted in their trisomy 21 (Down syndrome) children. From the work of Schupf and colleagues[37] discussed above, such mothers of Down syndrome children would also be expected to have an increased risk of developing Alzheimer's disease, again implicating chromosome mis-segregation in this disease process.

Further evidence for the potential involvement of cell cycle defects in Alzheimer's disease comes from the finding that both the amyloid precursor protein (APP) and the microtubule-associated protein tau found in Alzheimer paired helical filaments (PHF) become

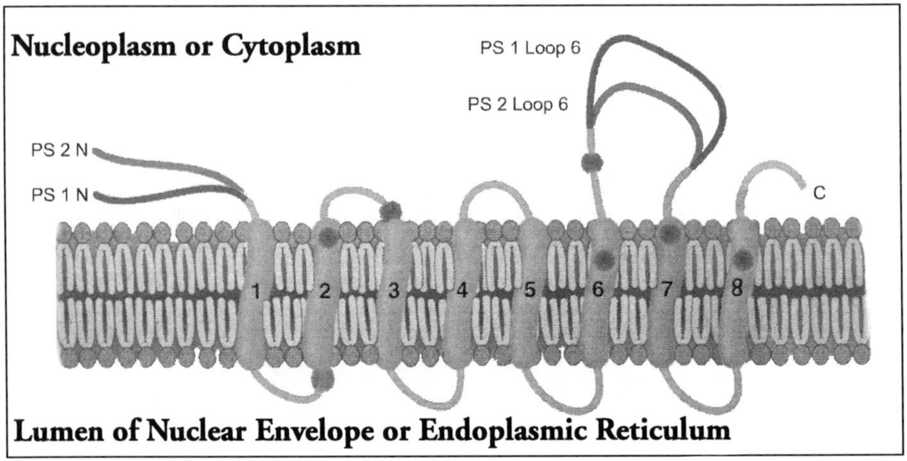

Figure 2. Probable transmembrane conformation of the two presenilin proteins (diagram modified by David Costa from ref. 188).

increasingly phosphorylated during mitosis[67-69] (Geller and Potter, unpublished). Thus, the two proteins most intimately involved in forming the characteristic neuropathological lesions of Alzheimer's disease are altered in structure, and presumably function, at different times in the cell cycle. Furthermore, phospho-tau and other mitosis-specific phospho-proteins and enzymes are present in neurons of Alzheimer, but not normal, brains.[70-77] Why these "post-mitotic" cells acquire mitosis-specific proteins is unknown. One possibility is that they have been stimulated to begin an aberrant mitosis. This, in turn, could have led to gain or loss of chromosomes, to changes in mitosis specific gene expression, or to apoptosis in response to an untenable physiological state—any or all of which could stimulate the development of Alzheimer pathology.[64,78,79]

Potential Mitotic Function of the Presenilin Proteins

As mentioned above, many of the Alzheimer individuals whose fibroblasts we analyzed for trisomy 21 mosaicism belonged to large families that were subsequently shown to carry familial Alzheimer's disease (FAD) mutations in two related genes (AD3 and AD4), now called presenilin 1 and 2 (Fig. 2). This result provided the first indication that familial Alzheimer's disease genes are likely to be involved in mitosis, and that their mutant forms predispose to chromosome mis-segregation—another of the predictions of the chromosome instability model of Alzheimer's disease.

The familial Alzheimer's disease genes encoding the presenilin proteins were identified on chromosome 14 (PS1) by St. George-Hyslop and colleagues based on previous genetic mapping[57,58] and on chromosome 1 (PS2) by Schellenberg, Tanzi and colleagues[59-61] and independently by us.[62] Both presenilin proteins include ~8 transmembrane domains and are structurally similar.

Almost forty point mutations have been identified throughout the two presenilin genes that cause early onset familial Alzheimer's disease (FAD). The fact that all of the FAD mutations reside within the coding regions of the genes suggests that a dominant gain of function resulting for instance in changes in their proteins' function, localization, or structure, rather than quantitative changes in expression, are likely to underlie the genes' involvement in Alzheimer's disease.

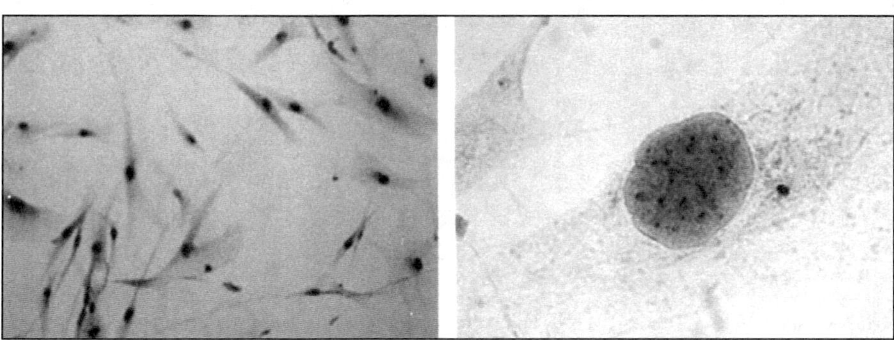

Figure 3. Primary human fibroblast cultures grown on cover slips were subjected to immunocytochemistry with affinity-purified antibodies to the presenilin loop region. Membrane, centrosome, and probable kinetochore labeling is evident. Reprinted with permission from Cell Press, ref. 78.

The presence of multiple transmembrane domains and potential target sites for the mitosis-specific cdc-2 kinase in the presenilins, together with the chromosome mis-segregation results with presenilin-mutant human cells, led us to suggest that the PS 1 and PS-2 proteins might reside in the nuclear membrane and might be involved in chromosome organization and segregation during the cell cycle.[62] We then tested this prediction by preparing a panel of antibodies to presenilin-derived peptides fused to glutathione-S-transferase and using immunocytochemical analysis to determine the normal intracellular location of the presenilin proteins.[78]

In the first approach to localizing the presenilins in the cell, PS-1- and PS-2-FLAG fusion proteins were expressed in transiently transfected cells and their intracellular location determined with anti-FLAG antibodies. Initial experiments indicated that the FLAG-tagged protein is partly localized in the endoplasmic reticulum and Golgi.[78] This was not un-expected because these are the sites of synthesis and processing of transmembrane proteins and because other groups had also overexpressed the presenilin proteins in cells and found them primarily localized to the endoplasmic reticulum.[81-83] However, because the localization of over-expressing proteins can easily become distorted, we sought to determine the subcellular location of the endogenous presenilin proteins. The affinity-purified presenilin antibodies were used to label the endogenous protein in cultured cells by Western blot analysis and immunocytochemistry.

Figure 3 shows an example of the intracellular localization of endogenous PS-2 in fibroblasts grown on cover slips. The labeling was primarily nuclear, which, in these flattened cells, was consistent with nuclear membrane localization. Putative PS-2 localization to the centrosome—identified as a small, strongly-labeled spot adjacent to the nucleus—was also apparent. These results, particularly the apparent centrosome labeling, provided additional evidence that the function of the PS-2 protein might be related to mitosis.

At higher magnification, the labeling of the nuclear membrane and the centrosome with the PS-2 loop antibody could be seen more clearly. In addition, a pronounced punctate pattern became apparent within the nucleus. This punctate nuclear membrane labeling pattern resembled the image generated by antisera directed against kinetochores—the protein complexes that bind the centromeres of the chromosomes and link them to the spindle microtubules during mitosis.[84,85] By adjusting the plane of focus, the PS-2-labeled spots could be seen to be located justaposed to the inner membrane of the nuclear envelope, where the kinetochores reside in interphase cells.[84,86]

To confirm the localization of PS-2 to the nuclear membrane, centrosomes, and kinetochores, we performed a series of double-label experiments with both the rabbit PS-2 loop

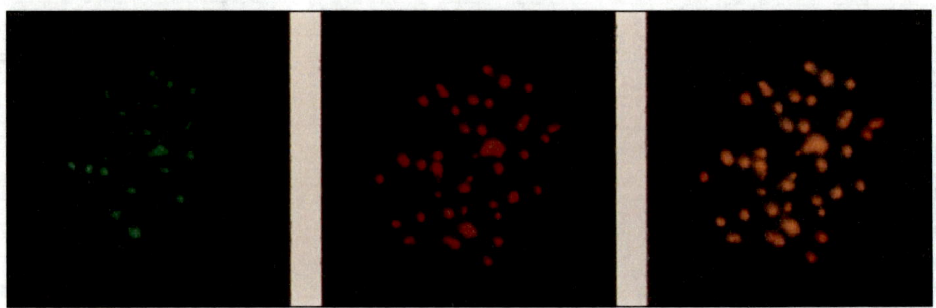

Figure 4. An example of a double label experiment performed using the presenilin 2 loop antibody and a CREST antibody previously shown to label kinetochores. Both antibodies stained the same punctate structures in the membrane of the cell. The right-most figure shows the co-labeling image of the FITC and CY3 secondary antibodies recognizing the presenilin 2 and kinetochore antigens. CREST antisera recognizes only kinetochore structures; the PS-2 antibodies also stain the nuclear membrane.

antibody and either a human anti-centrosome serum, a human anti-lamin serum, or a human anti-kinetochore serum, which could be distinguished from the PS-2 antibody with different fluorescent secondary antibodies.[78] For example, a human auto-immune serum was obtained from a patient with the CREST variant of scleroderma, shown previously to specifically label kinetochores in both metaphase and interphase cells[82,85] (F. McKeon, personal communication). The co-localization of the two signals (orange) identified the PS-2 protein as closely associated with the nuclear membrane, and with the kinetochores attached to the nuclear membrane during interphase (Fig. 4). Perhaps the membrane-associated presenilins serve as anchors to attach the kinetochores to the inner nuclear membrane during interphase. Consistent with this hypothesis is the finding that the centrosome labeling remains during mitosis, but the nuclear envelope labeling becomes dispersed and the kinetochores are unlabelled.[78]

Colocalization of staining with presenilin and centrosome antibodies is shown in Figure 5. Because the presenilins are clearly transmembrane proteins, we assume that the centrosome localization reflects the proteins association with cytoplasmic vesicles that have accumulated next to the centrosome.

Similar co-localization experiments were performed with an antibody to the N-terminus of PS-2 and also with two antibodies to the large hydrophilic loop and the N-terminus of PS-1.[78] Identical results were obtained with all four antibodies, indicating that both presenilins are associated with the nuclear membrane, centrosome, and kinetochores of normal dividing cells.

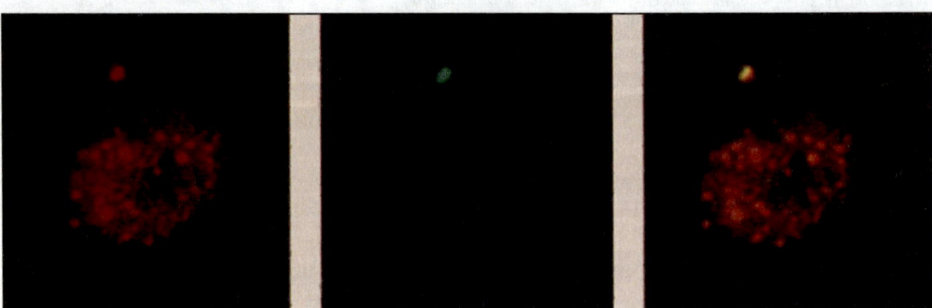

Figure 5. An example of a double immunolabeling of the centrosome with antibodies to presenilin 2 and to lamin.

Electron microscopy with gold-labelled secondary antibodies confirmed that the presenilins are associated with the inner nuclear membrane, the periphery of the centrosome, and punctate structures attached to the inner nuclear membrane.[78] Several other Alzheimer researchers have confirmed that a major location for the presenilins in the cell is the nuclear envelope[87,88] (B. Yankner, G. Schellenberg, T. Wisniewski and B. Frangione, B. De Strooper and G. Cole, personal communications). Furthermore, specific localization of presenilin protein to the centrosome and the kinetochore microtubules in early mouse embryos is induced by halting the cells in mitosis.[89] Finally, Johnsingh et al[90] and Pigino et al[91] have found that the presenilins associate with the cytoskeleton-associate proteins CLIP 170 and restin and with microtubules, consistent with our localization findings.

Functional results are also consistent with the presenilins being involved in mitosis. The FAD presenilin mutations have been found to inhibit the cell cycle[92] (Tanzi, personal communication; also see below), to increase sensitivity to apoptosis[93-95] (Ma and Potter, unpublished), and to prevent the full-length proteins from being translocated to the nuclear envelope.[88] Confirmation that the presenilin proteins are involved in chromosome segregation and Alzheimer's disease has recently been provided by the findings that a polymorphism in the presenilin 1 gene is associated with both an increased risk of developing Alzheimer's disease,[96-114] and also with an increased risk of having a Down syndrome child via a meiosis II defect.[115] Finally, when we examined the sequences of the PS1 and PS2, we found that in addition to the evident transmembrane domains, both proteins carry several S/TPXX amino acid motifs.[62] Based on their function in other proteins, these motifs may be target sites for cdc2 kinase and suggest that PS1 and PS2 reside in the nuclear membrane and are involved in mitosis. In vitro experiments on one of these sites confirm that it is an effective target for cdc2 kinase but not, for example for CKII (Li and Potter, unpublished).

The most straightforward interpretation of the subcellular localization of PS-1 and PS-2 in dividing cells is that presenilin function is related to mitosis, chromosome organization and segregation, and, potentially, to mitosis-specific gene expression. More specifically, the nuclear membrane and kinetochore localization of the presenilins suggests that one function of these proteins may be to serve as kinetochore binding proteins or receptors ("kinetoceptors") to anchor and organize the chromosomes on the nucleoplasmic surface of the inner nuclear membrane during interphase. In this role, the presenilins might bind DNA directly with special affinity for centromeric sequences, or they might interact with kinetochore proteins, possibly through microtubules. The point mutations and polymorphisms in the presenilin genes that cause or influence familial Alzheimer's disease may affect the ability of the presenilin proteins to link the chromosomes to the nuclear membrane and to release them at the appropriate time during mitosis, thus leading to chromosome mis-segregation and other consequent abnormalities seen in cells carrying mutant presenilin genes, such as inappropriate apoptosis. Interestingly, there are precedents for nuclear membrane proteins having a role in mitosis. Specifically, the yeast NDC-1 protein and the lamin B receptor both have seven-eight transmembrane domains, are located in the nuclear membrane, contain phosphorylatable S/TPXX motifs, and, in the case of NDC-1, cause chromosome mis-segregation when mutant.[62,116-118]

In order for a cell to leave G1 and enter mitosis, the kinetochores and their associated chromosomes must be released from the surface of the inner nuclear membrane and its associated proteins such as the presenilins. Two possible ways in which the putative presenilin binding to centromeres/kinetochores might be regulated during this process are by phosphorylation or by proteolytic cleavage. It is therefore interesting that PS-2 is phosphorylated in vivo by casein kinase 2,[82] and in vitro by the cdc-2 kinase[62] (Li and Potter, in preparation), and that both presenilin proteins, though found in full-length form, are also cleaved into two parts in normal and transfected cells.[78,119,120]

Table 1. Transient transfection experiments

DNA	Abnormal	Karyotypes Total	% Abnormal
none	17	237	7.2
Vector	10	143	7.0
PS-1	7	19	36.8***
PS-2	23	170	13.5*

Lymphoblastoid cells were transiently transfected with the indicated vectors, allowing the expression of different presenilin proteins. Karyotype analysis was carried out 72 hours after transfection and the number of abnormal karyotypes counted.

Mitotic Effects of the Normal and FAD Mutant Presenilin Proteins in Transfected Cells

The evidence that the presenilins are involved in mitosis and chromosome segregation is provided by their location within the cell and the finding that fibroblasts from individuals harboring an FAD mutation in PS-1 or PS-2 show chromosome mis-segregation. To further investigate possible mitotic functions of the PS proteins, the normal and FAD mutant versions of their genes have been cloned into expression vectors and introduced into mammalian cells with normal chromosome complements.

Preliminary evidence from transfected cells over-expressing the presenilin proteins are consistent with their involvement in chromosome segregation. AG09393 cells were transiently transfected with the pcDNA-3 vector either unmodified or expressing wild type PS-1 or PS-2. The combined level of aneuploidy in several experiments was statistically significantly higher in the cells receiving the PS-containing vectors compared to the cells receiving water or the vector (pcDNA3) alone (see Table 1). Over-expression of even normal presenilins clearly affects chromosome segregation. The chromosome changes included trisomies, monosomies, and translocations.

The fact that transient over-expression of even normal presenilin proteins causes massive chromosome instability and apoptosis,[93,94] suggested that a more controlled expression system would be better able to show differences between the wild type and mutant presenilin genes. To this end, we have developed tetracycline activated expression vectors and are in the process of analyzing their effect on chromosome stability.

Chromosome and Cell Cycle Changes in Cells from Presenilin Transgenic and Knockout Mice

The demonstration that the presenilin proteins in mice were located in the same apparent structures as in human[62] prompted some preliminary experiments in mouse cells to determine the effect of PS-1 deletion on the cell cycle. Shen et al developed a PS-1 knockout mouse which could be maintained in the heterozygous state and could generate homozygous knockout embryos.[121] We prepared primary fibroblast cultures from such embryos and measured the cell division time and the number of in vitro passages needed to reach senescence compared to cells from a normal littermate. PS-1 knockout cells clearly grew more slowly than normal cells, and many of the knockout cells reached senescence as evidenced by flattening and cessation of mitosis, before any of the normal cells did.

Table 2. Aneuploidy in PS1 knockout liver

	# of Cells	
Chrm #	wt	PS1-/-
40 (normal)	68	56
80	0	1
39	1	2
37	0	1
27	0	1
< 20	4	8
% abnormal	6.8	18.8*
Mitotic Index	50%	20%

Liver cells from wild type and PS-1 knockout embryos were grown in culture and their chromosomes prepared for karyotype analysis. The knockout cells grew more slowly and developed increased aneuploidy.

We have also carried out a preliminary experiment to analyze the chromosome complement of fetal liver cells from a PS-1 knockout embryo in comparison to a normal littermate. The number of karyotypes that were abnormal was substantially (and statistically significant; P<0.05) higher in the E-15 PS-1 null liver cells than in cells obtained from a normal wild type littermate (see Table 2). In addition, the mitotic index was substantially smaller in the PS-1 KO cell cultures, also suggesting the presence of a defect in the cell cycle.

The results from the PS-1 knockout mouse cells have been complemented by similar experiments on cells derived from transgenic mice harboring and expressing the human PS-1 normal and PS-1-FAD mutant genes under the control of the PDGF promoter.[122] Thus far, we have karyotype data from hundreds of spleen cells (in which the PDGF promoter is active) that reveal substantial aneuploidy in mice carrying a PS-1 gene harboring the M146L FAD mutation (Table 3). The number of aneuploid cells increases with the age of the transgenic mice. Flow cytometry confirms these results and also shows no aneuploidy in wt PS mice of 17 and 24 months of age.

All of these results are consistent with PS-1 playing an important role in the normal physiology of dividing cells. The chromosome analysis specifically suggests that normal amounts of WT presenilins are essential for maintaining the fidelity of chromosome segregation. This

Table 3. Aneuploidy in PS-1 mutant mice

	Chromosome Complement		
Strain (Age)	40 (Normal)	Aneuploid	% Aneuploid
PS1 M146L (8 Mo)	558	9	1.6**
PS1 M146L (13 Mo)	489	16	3.2**
PS1 wt (16 Mo)	70	0	0
Non Tg (7Mo)	194	0	0

Spleen cells from mice carrying the wild type or a mutant presenilin transgene and from nontransgenic mice were stimulated to divide by ConA and their chromosomes analyzed following colcemid treatment 44 hours later.

result, if confirmed, could also explain the slower growth, the reduced lifespan in culture, and the tendency to undergo apoptosis of cells either knocked out or mutant for presenilins. The result also confirms the previous finding of chromosome mis-segregation and trisomy 21 in fibroblasts from presenilin mutant individuals with Alzheimer's disease.[55]

In Alzheimer's disease research, it has become almost axiomatic that the main cause of the disease is an increase in the relative amounts of the Alzheimer Aβ 1-42 and 1-40 peptides, with the Aβ 1-42 peptide being more prone to form amyloid deposits and induce neuronal damage.[4] However, it is worth pointing out that the many, many fold increase in chromosome aneuploidy in cells carrying a presenilin mutation far exceeds the barely 2-fold effect of the mutations on the relative production of Aβ 1-42 and 1-40. Perhaps the main effect of the presenilin mutations is on the cell cycle and chromosome segregation, with a coincidental side effect being an alteration in Aβ production.

Mutation/Inactivation of the Tumor Suppressor p53 Causes Chromosome Instability and Regulates Expression of Presenilin

A completely independent result from the study of cancer supports the conclusion that the presenilins are involved in mitosis and chromosome segregation. p53 is a well-studied tumor suppressor gene whose mode of action is still not completely understood. Lack of p53 clearly predisposes to malignancy in humans and experimental animals. Indeed, p53 is the most common mutation in human cancers, and animals, humans, and cell lines mutant for p53 show abnormal centrosome duplication, chromosome mis-segregation, and aneuploidy.[123-126] Thus p53 may exert its tumor suppressor function at least partly by promoting the orderly segregation of chromosomes during mitosis, with loss of p53 leading to aneuploidy. This conclusion is supported by two recent findings that together link p53 function to chromosome segregation and to the presenilins.

One approach to determining how a tumor suppressor gene acts to suppress unrestricted cell division is to identify other genes whose expression is either up-regulated or down-regulated by the tumor suppressor gene product and then to determine which are important downstream effectors of the pathway. One such study used a p53 mutant cell line into which a vector construct that expressed a temperature sensitive p53Val135 protein was stably transfected.[127] Shifting the cells between the permissive and nonpermissive temperature and carrying out differential RNA display identified a series of genes that are up-regulated or down-regulated by wt p53. One such gene that was down-regulated is presenilin 1 (PS1), previously identified as carrying multiple mutations that cause autosomal dominant Alzheimer's disease.

Roperch and colleagues then assessed the importance of PS1 in p53 function by knocking out the PS1 protein by transfection of p53 mutant cells (which express large amounts of PS1) with an PS1 antisense RNA-expressing construct.[127] The cells grew much more slowly, lost their ability to be cloned in soft agar, and generated few tumors when injected into nude mice. Those tumors that did develop had regained their presenilin expression, confirming the conclusion that PS1 is essential for the tumor-causing effect of the loss of p53 function.

The data from the Alzheimer's disease patients fibroblasts and the transgenic mice together with the suggestive involvement with p53 provide a strong indication that the presenilins may be involved in chromosome segregation, The location of the presenilins at the interphase kinetochore suggest a mechanism by which mutations in the presenilins could cause chromosome instability and aneuploidy. Appropriate chromosome segregation requires that very careful control of chromosome location and organization be maintained throughout the cell cycle, not merely during mitosis. This is presumably the reason that the kinetochores are attached to the inner surface of the nuclear membrane during interphase rather than being allowed to roam freely in the nucleoplasm. Therefore, it is likely that if this interphase location were to be disrupted by mutation in a protein such as a presenilin that seems to be connecting the interphase kinetochores with the inner nuclear membrane, then the end result could be a defect in chromosome segregation. For instance, if the chromosomes do not detach from the nuclear membrane at the right time and with the right speed and coordination, then they may not become available for attachment by the spindle microtubules with equal efficiency, and lagging chromosomes may become mis-segregated.

Presenilin Mutations May Effect Mitosis in both Neurons and Glia

The presence of trisomy 21 mosaicism in skin fibroblasts from Alzheimer's disease patients with presenilin mutations suggests that other cells from these individuals are also likely to be mosaic for trisomy 21. Which cells would, when aneuploid for chromosome 21, contribute most to Alzheimer neuropathology? The answer is not obvious. Neurons are the cells whose loss leads to the clinical symptoms of Alzheimer's disease and neurons produce large amounts of APP and of Aβ. Because mature neurons do not divide, it might be thought that they would be unaffected by mitotic proteins. There are two potential ways that FAD mutations in the presenilins could affect neurons directly. First, the presenilins may have an additional, non-nuclear, function in neurons, and, indeed, the majority of the presenilin protein in these cells is present in the somato-dendritic compartment. This localization has led to the reasonable hypothesis that FAD mutations in the presenilins change their interaction with the APP protein and consequently alter APP processing to produce the increased Aβ$_{1-42}$ seen in presenilin mutation-carrying patients, cells, and transgenic mice. However, the results reviewed here force us to consider whether the apparent involvement of the presenilins in mitosis might also affect neuronal function and Aβ production. For example, it has become clear that new neurons are, in fact generated in the hippocampus of higher animals, including primates, not only during development, but throughout adulthood as well.[128-138] Specifically, the rate of new neuron formation in the dentate gyrus of the hippocampus is reported to be on the order of one to two thousand cells per day—more than enough to cause significant numbers of trisomy 21 or other aneuploid cells to accumulate over the course of 50 years under the influence of, for instance, a presenilin mutation. Alternatively, neurons may respond to mutations in FAD genes such as the presenilins by attempting to undergo inappropriate division, and perhaps becoming aneuploid and then undergoing apoptosis in the process. Indeed the aneuploid cells reported in AD patients by Yang and colleagues were brain neurons.[64] Significantly, as mentioned above, phospho-epitopes and enzymes normally observed only in mitotic cells have been found in neurons in Alzheimer's disease brain. Furthermore, apoptotic neurons generate and release more Aβ than do healthy neurons.[139,140]

To an even greater extent than neurons, the glial cell population continues to divide in the adult primate brain[141] and therefore could accumulate aneuploid cells throughout life. The involvement of glial cells in the neuropathology of Alzheimer's disease is indicated by increasing evidence that these cells mount an inflammatory reaction and an acute phase response that make an important contribution to the characteristic pathology and neuronal cell death of the Alzheimer's disease brain.[7,142-149] Particularly interesting is the fact that an abnormally high number of microglia overexpress the lymphokine IL-1 in the areas of Alzheimer's disease and Down syndrome brain that exhibit, or will exhibit, Alzheimer's disease neuropathology.[148,150,151] Il-1, in turn, induces astrocytes to express the amyloid-associated protein α_1-antichymotrypsin (ACT). Both ACT and another amyloid-associated protein elicited by inflammation, apo E, can facilitate the polymerization of Aβ peptide into neurotoxic amyloid filaments.[6-8,148,149,152] The essential function of apoE and ACT as amyloid promoting factors in vivo has been demonstrated recently in animal models of Alzheimer's disease.[9-12] Indeed, we have recently shown that product or process of ACT and apoE-catalyzed amyloid formation in transgenic mice is responsible for their cognitive decline.[12] In sum, a glial cell-led inflammatory cascade appears to play an essential role in Alzheimer amyloid formation, starting with the release of Il-1 from microglia. The constitutive activation of microglia in Down syndrome brain, even before birth, suggests the possibility that if microglia in a normal individual acquire an extra copy of chromosome 21, they may release Il-1 and initiate the inflammatory cascade that leads to the neuropathology of Alzheimer's disease.

Although it would seem reasonable that amyloid should develop in the regions immediately surrounding aberrant cells (for instance trisomy 21 neurons or glial cells), the precedent provided by other amyloidoses suggests that this need not be the case. For instance, the autosomal-dominantly inherited diseases Familial Amyloidotic Polyneuropathy and Hereditary Cerebral Hemorrhage with Amyloidosis of both the Dutch and Icelandic types have very specific regions of amyloid deposition despite the fact that all cells in the body carry the point mutation in the affected amyloid gene (transthyretin, cystatin C, or APP respectively), and that these genes are expressed in many parts of the body where the amyloid does not deposit.[153] Thus, by analogy, the trisomy 21 cells that are relevant for the formation of amyloid pathology in Down syndrome (and, according to the hypothesis presented here, Alzheimer's disease) need not reside in the brain at all. Indeed, some researchers have suggested that the $A\beta$ peptide is transported to the brain by the circulation, after having been generated elsewhere.[154]

How Aneuploidy May Lead to Alzheimer's Disease

Several potential mechanisms could explain how the generation of trisomy 21 or other aneuploid cells by FAD-mutant presenilin proteins could lead to Alzheimer's disease. For example, aneuploid cells might be prone to apoptosis. Indeed, cortical neurons from Down syndrome fetuses undergo spontaneous apoptosis in vitro.[155] Such apoptosis would lead to neurodegeneration directly, but it could also indirectly affect APP processing and the production of the $A\beta$ peptide. Support for this latter hypothesis is provided by the findings that Down syndrome fetal brains and adult sera contain a higher ratio of the pathogenic $A\beta_{1-42}$ compared to $A\beta_{1-40}$[156] and that inducing apoptosis in normal human neurons by serum starvation or other treatments increases their secretion of the $A\beta$ peptide.[139,140] A more direct connection between the presenilin genes and apoptosis is indicated by the finding that the overexpression of even normal presenilin genes in transfected cells (which we would expect to disrupt orderly chromosome segregation) results in increased sensitivity of the cells to apoptotic stimuli[95,157,158] (Ma and Potter, unpublished). Furthermore, expression of a C-terminal fragment of PS-2 in cells was shown to inhibit induced apoptosis, which was then restored by overexpression of full-length PS-2.[93,94]

How could presenilin-induced apoptosis lead to the other commonly-reported effect of expressing the FAD mutant presenilins in human patients, tranfected cells, or trangenic animals—an increase in the ratio of secreted $A\beta_{1-42}$ over $A\beta_{1-40}$?[122,159-162] One possibility is that the mutant presenilins induce changes in mitosis and chromosome segregation that affect the level of mitosis-specific phosphorylation of various proteins. In as much as altering the phosphorylation of APP alters the production and secretion of $A\beta$, and the phosphorylation of the APP protein naturally changes during the cell cycle,[67] a cell cycle-altering mutation in a protein such as one of the presenilins could affect APP processing indirectly through its phosphorylation state and thus lead to the observed increase in $A\beta_{1-42}$.

Potential Involvement of Other Genes in Chromosome Mis-Segregation in Alzheimer's Disease

Thus far, it is clear that trisomy 21 mosaicism can cause some sporadic cases of Alzheimer's disease, and there is strong circumstantial evidence favoring such a mechanism as one explanation for how the FAD mutations in the presenilin genes exert their Alzheimer-causing effect. Thus a key prediction of the trisomy 21 mosaicism model for Alzheimer's disease has been, at least partly, fulfilled. Is it possible that other forms of Alzheimer's disease also involve trisomy 21 mosaicism? Here the evidence is intriguing. For example, FAD mutations in the APP gene result both in apoptosis and in increased production of $A\beta_{1-42}$ over $A\beta_{1-40}$, which is also characteristic of trisomy 21 and of the presenilin mutations.[163-166] Then there is the finding that the APP protein not only becomes hyperphosphorylated during mitosis,[67] but that APP phosphorylation also modulates APP processing.[167-171] Furthermore, endogenous APP localizes to the centrosome and the nuclear membrane in dividing cells just as we have found for the presenilins[172,173] (Li and Potter, unpublished), and has also been found to interact with the presenilins by co-immunoprecipitation from extracts of APP and PS2 double-overexpressing cells.[174] Also, in our study of aneuploidy in Alzheimer's disease cell lines, the two lines harboring an APP mutation also exhibited increased trisomy 21.[56] Perhaps the APP mutations act not directly, but indirectly, on APP processing, by altering the cell cycle and its pattern of phosphorylation of proteins including APP itself. For example, the cytoplasmic tail of APP binds to a protein that drives cells though the S-M checkpoint and causes apoptosis in neurons.[175] Finally, knocking a gene that is highly homologous to APP, termed APLP2, in mice is lethal because the cells of the very early embryo undergo massive chromosome mis-segregation.[176,177]

The other major genetic risk factor for Alzheimer's disease is the inheritance of the ε4 allele of apolipoprotein E. Although the best evidence indicates that the apoE4 protein promotes Alzheimer's disease by promoting the polymerization of the Aβ peptide into neurotoxic amyloid filaments (discussed above), it is possible that apoE4 has additional functions related to mitosis and chromosome segregation. Such a conclusion was reached, for example, by Avramopoulos and colleagues who showed that young mothers of Down syndrome children are significantly more likely to carry an apo ε4/ε4 genotype.[178] The authors suggested that the demonstrated inability of the apoE4 protein (compared to apoE3) to bind to the microtubule-associated protein tau[179] disrupts the balance of microtubule polymerization/depolymerization, and thus affects the mechanics of chromosome segregation. This explanation is interesting in light of the facts that a similar group of Down syndrome mothers was shown by Schupf et al to have an increased risk of developing Alzheimer's disease[37] and that the highest level of trisomy 21 mosaicism in our sample of Alzheimer's disease patients was found in an apo ε4/ε4 individual.[56] It is also interesting that exogenously-added apoE halts lymphocyte proliferation in the G1A phase of the cell cycle.[180] Further biochemical and chromosome segregation studies are underway to test directly whether apoE4 affects chromosome segregation.

Although improper chromosome segregation can result from a genetic mutation, it can also be caused by environmental agents. Of the many exogenous factors that influence chromosome segregation, microtubule-disrupting agents such as colchicine and low doses of radiation are perhaps the best studied.[181] Aluminum, the consumption of which some studies show to have a weak but significant association with the development of Alzheimer's disease, also binds to microtubules and, in the form of aluminum silicate, causes chromosome nondisjunction in cultured cells.[182-184] Thus, the large proportion of Alzheimer's disease cases that arise in a sporadic manner not directly attributable to the inheritance of a genetic mutation can also be understood in the light of the chromosome 21 trisomy model.

Implications of Trisomy 21 Mosaicism in Alzheimer's Disease for Future Diagnosis and Therapy

The mechanistic implication of the findings discussed above is that an early step in the pathogenic pathway to Alzheimer's disease may involve a defect in chromosome segregation that leads to trisomy 21 mosaicism. This conclusion can be exploited in our search for more effective diagnoses and treatments for Alzheimers disease.

As already discussed, the similar hypersensitivity of Alzheimer and Down syndrome individuals to cholinergic agonists or antagonists may serve as an Alzheimer diagnostic test. Another, very straightforward potential diagnostic test based on our finding of trisomy 21 cells among Alzheimer patients fibroblasts would be to use chromosome analysis to directly assess the level of trisomy 21 mosaicism in an individual. Although we have found that cell lines from Alzheimer individuals as a group have increased trisomy 21 mosaicism, there is substantial overlap in the data between the Alzheimer patients and the control individuals. Clearly, the test must be made more sensitive to be useful. One approach, which we have initiated, is to use primary cells directly derived from the patients rather than fibroblast cell lines. This would reduce the influence of culture conditions on the chromosome analysis and potentially provide more rapid results. Specifically, buccal cells, which the individual scrapes with a soft brush from the inside of their cheek, can be assayed for trisomy 21 and other aneuploidy by modification of the standard fluorescence in situ hybridization (FISH) technique. This scraping method also has the advantage of sampling cells from a larger area than does a needle skin biopsy, thus reducing the possibility of missing a small clone of trisomy 21 or other aneuploid cells that might be important. It can also be repeated a number of times to strengthen the confidence of the analysis. The data thus far are too preliminary to make strong conclusions, but it is reasonable to be concerned that an individual who is clearly and reproducibly found to be mosaic for trisomy 21 may be at risk for developing Alzheimer's disease later in life. Similarly, as discussed above, women who have a trisomy 21 fetus at an early age might consider undergoing chromosome analysis of not only their blood lymphocytes, but also their fibroblasts, to assess whether or not trisomy 21 mosaicism is present.

The development in the last ten years of a clearer understanding of the pathogenic pathway that leads to Alzheimer's disease has identified key steps in this pathway that might serve as targets for therapeutic intervention. For instance, the APP processing enzymes that generate the Aβ peptide, the process of Aβ polymerization itself, the amyloid-inducing interaction of ACT or apoE with Aβ, the production of IL-1 by microglial cells, or its recognition by the receptors on the surfaces of astrocytes are all potential steps in the pathway that our current knowledge suggests would be effective Alzheimer therapeutic targets, and for which drug screening assays can be designed. The implication of the data presented in this review that trisomy 21 mosaicism may be one of the very first steps in the Alzheimer pathogenic pathway suggests a new set of approaches to therapy that may be particularly valuable because they are targeted at the very earliest step in the pathway. In some situations it may be possible to determine the mechanism of chromosome mis-segregation or defective mitosis-specific gene expression. Drugs that would rectify the defect could then be selected for. For example, if the familial Alzheimer's disease (FAD) mutations in the presenilins result in a kinetochore receptor that fails to bind the kinetochore sufficiently tightly to prevent chromosomes from becoming disorganized at the start of mitosis, a drug that improved presenilin kinetochore binding could be developed. Similarly, if apoE4 interaction with microtubules is defective and leads to chromosome mis-segregation, a drug could be developed to increase the effectiveness of apoE4 in individuals carrying this allele.

Alternatively, it may be possible to strengthen the fidelity of the chromosome segregation process prophylactically. Several approaches are possible. It may be possible to identify and eliminate environmental toxins that cause chromosome mis-segregation, or to develop agents that counteract the effect of the toxins. Such treatments might include heavy metal chelaters, antioxidants, and promoters of microtubule assembly. Drugs that improve chromosome segregation might also include those that affect DNA topoisomerase II or centromere binding proteins such as CBF1 or DIS1.[185-187]

A more difficult, but potentially equally effective approach to therapy, would be to determine whether it is the aneuploid—for instance, trisomy 21—cells per se that are contributing to the development of Alzheimer's disease, as appears to be the case in Down's syndrome, or whether it is the secondary effects of aberrant mitosis or attempted mitosis that are important. If the latter is the case, then the drugs discussed above would be appropriate. If the trisomy 21 cells themselves are the problem, then it may be possible to target them for removal from the body by exploiting unusual features of their cell biology. Specifically, several genes are overexpressed in trisomy 21 cells to a much greater extent than would be expected by the 50% increase in gene dosage. An example is the APP protein itself; another is the interferon α-2 receptor. Antibodies that recognize such cell surface antigens that are overexpressed to a great extent in trisomy 21 cells could be coupled to cellular toxins and used to selectively eliminate these cells. Similarly, if the preliminary evidence indicating that trisomy 21 microglial cells are constitutively active and expressing the lymphokine IL-1 is confirmed, it may be possible to eliminate IL-1 positive microglial cells at a single point in time by microglia-targeted gene therapy with a toxin gene under the control of a regulatable IL-1 promoter. Trisomy 21 microglial cells that receive the gene would self-destruct, but only if another co-inducting signal were also present—for instance, tetracycline. Once the trisomy 21 cells were eliminated, the co-inducer could be removed, and the remaining, diploid, microglial cells (though still containing the toxin gene under the control of the IL-1 promoter and the tetracycline enhancer) would still be able to carry out their normal function during infection or other demands for brain inflammation in which the IL-1 cytokine must be produced under proper regulatory control. Because the toxin gene would not be activated, the normal microglial cells could be induced to express IL-1 without being killed.

In summary, some Alzheimer's disease patients are clearly mosaic for trisomy 21, and many others may have low levels of trisomy 21 mosaicism. The mechanism(s) by which these abnormal cells arise and whether/how they contribute to the pathogenesis of the disease are areas of active investigation. Further exploration of these novel findings has the potential to contribute to the development of diagnoses and therapies for Alzheimer's disease and to increase our understanding of basic cellular processes.

References

1. Potter H, Ma J, Das S et al. Beyond β-protein: new steps in the pathogenic pathway to Alzheimer's disease. In: Iqbal K, Mortimer JA, Winblad B et al, eds. Research Advances in Alzheimer's Disease and Related Disorders. New York: John Wiley and Sons, 1995:643-654.
2. Yankner BA. Mechanisms of neuronal degeneration in Alzheimer's disease. Neuron 1996; 16:921-932.
3. Potter H. Beyond beta protein—The essential role of inflammation in Alzheimer amyloid formation. In: Bondy SC, Campell A, eds. Inflammatory Events in Neurodegeneration. Prominent Press, 2001.
4. Selkoe DJ. Alzheimer's disease: Genes, proteins, and therapy. Physiol Rev 2001; 81:741-766
5. Potter H, Wefes IM, Nilsson LNG. The inflammation-induced pathological chaperones ACT and apoE are necessary catalysts of Alzheimer amyloid formation. Neurobiol Aging 2001; in press.
6. Ma J, Yee A, Brewer HB et al. The Alzheimer amyloid-associated proteins α_1-antichymotrypsin and apolipoprotein E promote the assembly of the Alzheimer β-protein into filaments. Nature 1994; 372:92-94 (see also "News and Views," same issue).
7. Wisniewski T, Castaño EM, Golabek A et al. Acceleration of Alzheimer's fibril formation by apolipoprotein E in vitro. Am J Pathol 1994; 145:1030-1035.
8. Bales KR, Verina T, Dodel RC et al. Lack of apolipoprotein E dramatically reduces amyloid β-peptide deposition. Nature Genetics 1997; 17:263-264.
9. Holzman DM, Bales KR, Tenkova T et al. Apolipoprotein E isoform-dependent amyloid deposition and neuritic degeneration in a mouse model of Alzheimer's disease. Proc Natl Acad Sci USA 2000; 97:2892-2897.
10. Mucke L, Yu G-Q, McConlogue L et al. Astroglial expression of human α_1-antichymotrypsin enhances Alzheimer-like pathology in amyloid protein precursor transgenic mice. Am J Pathol 2000; 157:2003-2010.

11. Nilsson LNG, Bales KR, DiCarlo D et al. α-1-antichymotrypsin promotes beta-sheet amyloid plaque deposition in transgenic mouse model of Alzheimer's disease. J Neurosci 2001; 21:1444-1451

12. Nilsson LNG, Arendash GW, Low MA et al. Cognitive impairment depends on ApoE and ACT-catalyzed amyloid formation in an Alzheimer's mouse model. (submitted)

13. Olson MI, Shaw CM. Presenile dementia and Alzheimer's disease in mongolism. Brain 1969; 92:147-156.

14. Glenner G, Wong CW. Alzheimer disease and Down's syndrome: Sharing of a unique cerebrovascular amyloid fibril protein. Biochem Biophys Res Commun 1984; 122:1131-1135.

15. Wisniewski HM, Rabe A, Wisniewski KE. Neuropathology and dementia in people with Down's syndrome. In: Davies P, Finch C, eds. Molecular Neuropathology of Aging. New York: Cold Spring Harbor Laboratory, 1988:399-413.

16. Epstein CJ. The consequences of chromosome imbalance. Am J Med Genet Supp 1990; 7:31-37.

17. Schweber M. A possible unitary genetic hypothesis for Alzheimer's disease and Down syndrome. Ann NY Acad Sci 1985; 450:223-238.

18. Schweber MS. Alzheimer's Disease Related Disorders. San Diego: Alan R Liss, Inc. 1989:247-267.

19. Goldgaber D, Lerman MJ, McBride OW et al. Characterization and chromosomal localization of a cDNA encoding brain amyloid of Alzheimer's disease. Science 1987; 235:877-880.

20. Kang J, Lemaire HG, Unterback A et al. The precursor of Alzheimer disease amyloid A4 protein resembles a cell-surface receptor. Nature 1987; 325:733-736.

21. Robakis NK, Ramakrishna N, Wolfe G et al. Molecular cloning and characterization of a cDNA encoding the cerebrovascular and the neuritic plaque amyloid peptides. Proc Natl Acad Sci USA 1987; 84:4190-4194.

22. Tanzi RE, Gusella JF, Watkins PC et al. Amyloid β-protein gene; cDNA, mRNA distributions, and genetic linkage near the Alzheimer locus. Science 1987; 235:880-884.

23. Neve RL, Finch EA, Dawes LR. Expression of the Alzheimer amyloid precursor gene transcripts in the human brain. Neuron 1988; 1:669-677.

24. Rumble B, Retallack R, Hilbich C et al. Amyloid A4 protein and its precursor in Down's syndrome and Alzheimer's disease. NEJM 1989; 320(22):1446-1452

25. Oyama F, Cairns NJ, Shimada H et al. Down's syndrome: Up-regulation of beta-amyloid protein precursor and tau mRNAs and their defective coordination. J Neurochem 1994; 62(3):1062-1066.

26. Potter H. Review and hypothesis: Alzheimer disease and Down syndrome—Chromosome 21 non-disjunction may underlie both disorders. Am J Hum Genet 1991; 48:1192-1200.

27. Scinto LFM, Daffner KR, Dressler D et al. A potential non-invasive neurobiological test for Alzheimer's disease. Science 1994; 266:1051-1054.

28. Scinto LFM, Rentz DM, Potter H et al. Pupil assay and Alzheimer's disease: A critical analysis. Neurology 1999; 52:673.

29. Scinto LMF, Wu CK, Firla KM et al. Focal Pathology in the Edinger-Westphal nucleus explains pupillar hypersensitivity in Alzheimer's disease. Acta Neuropath 1999; 97:557-564.

30. Scinto LF, Frosch M, Wu CK et al. Selective cell loss in Edinger-Westphal in asymptomatic elders and Alzheimer's patients. Neurobiol Aging 2001; 22:729-36.

31. Heston LL, Mastri AR. The genetics of Alzheimer's disease. Arch Gen Psychiatry 1977; 34:976-981.

32. Heston LL, Mastri AR, Anderson VE et al. Dementia of the Alzheimer type. Arch Gen Psychiat 1981; 38:1084-1090.

33. Heyman A, Wilkinson W, Hurwitz B et al. Alzheimer's disease: Genetic aspects and associated clinical disorders. Ann Neurol 1983; 14:507-515.

34. Whalley LJ, Carothers AD, Collyer S et al. A study of familial factors in Alzheimer's disease. Brit J Psychiat 1982; 140:249-256.

35. Amaducci LA, Fratiglioni L, Rocca WA et al. Risk factors for clinically diagnosed Alzheimer's disease: A case-control study of an Italian population. Neurology 1986; 36:922-931.

36. Chandra V, Philipose V, Bell PA et al. Case-control study of late onset "probable Alzheimer's disease." Neurology 1987; 37:1295-1300.

37. Schupf N, Kapell D, Lee JH et al. Increased risk of Alzheimer's disease in mothers of adults with Down's syndrome. The Lancet 1994; 344:353-356.

38. Schapiro MB, Kumar A, White B et al. Alzheimer's disease (AD) in mosaic/translocation Down's syndrome (Ds) without mental retardation. Neurology 1989; 39:169.

39. Rowe IF, Ridler MAC, Gibberd FB. Presenile dementia associated with mosaic trisomy 21 in a patient with a Down syndrome child. Lancet 1989; July:229.

40. Hardy J, Goate A, Owen M et al. Presenile dementia associated with mosaic trisomy 21 in a patient with a Down syndrome child. Lancet 1989; September:743

41. Percy ME, Markovic VD, McLachlan DRC et al. Family with 22-derived marker chromosome and late-onset dementia of the Alzheimer type: I. Application of a new model for estimation of the risk of disease associated with the marker. Am J Med Genet 1991; 39:307-313.

42. Jarvik LF, Yen F-S, Goldstein F. Chromosomes and mental status. A study of women residing in institutions for the elderly. Arch Gen Psychiat 1974; 30:186-190.

43. Ward BE, Cook RH, Robinson A et al. Increased aneuploidy in Alzheimer disease. Am J Med Genet 1979; 3:137-144.

44. Nordenson I, Adolfsson R, Beckman G et al. Chromosomal abnormality in dementia of Alzheimer type. Lancet 1980; 1:481-482.

45. White BJ, Crandall C, Goudsmit J et al. Cytogenetic studies of familial and sporadic Alzheimer disease. Am J Med Genet 1981; 10:77-89.

46. Buckton KE, Whalley LJ, Lee M. Chromosome changes in Alzheimer's presenile dementia. J Med Genet 1983; 20:46-51.

47. Moorhead PS, Heyman A. Chromosome studies of patients with Alzheimer's disease. Am J Med Genet 1983; 14:545-556.

48. Kormann-Bortolotto MH, Smith MAC, Neto JT. Alzheimer's disease and ageing: A chromosomal approach. Gerontol 1993; 39:1-6.

49. Lau JT, Beyer K, Fernadez-Novoa L et al. Cytogenetic signs of genomic instability in Alzheimer's disease. Neurobio Aging 1996; 17:S62 (supplement).

50. Migliore L, Testa A, Scarpato R et al. Spontaneous and induced aneuploidy in peripheral blood lymphocytes of patients with Alzheimer's disease. Hum Genet 1997; 101:299-305.

51. Fitzgerald PH, Pickering AF, Mercer JM et al. Premature centromere division: A mechanism of non-disjunction causing X chromosome aneuploidy in somatic cells of man. Ann Hum Genet 1975; 38:417-428

52. Fitzgerald PH, Archer SA, Morris CM. Evidence for the repeated primary non-disjunction of chromosome 21 as a result of premature centromere division (PCD). Hum Gene 1986; 72:58-62.

53. Lichter P, Cremer T, Tang CJC et al. Rapid detection of human chromosome 21 aberrations by in situ hybridization. Proc Natl Acad Sci USA 1988; 85:9664-9668.

54. Fuscoe JC, Collins CC, Pinkel D et al. An efficient method for selecting unique-sequence clones from DNA libraries and its application to fluorescent staining of human chromosome 21 using in situ hybridization. Genomics 1989; 5:100-109

55. Potter H, Geller LN. Alzheimer's disease, Down's syndrome, and chromosome segregation. The Lancet 1996; 348:66.

56. Geller LN, Potter H. Chromosome mis-segregation and trisomy 21 mosaicism in Alzheimer's disease. Neurobiol Disease 1999; 6:167-179.

57. Schellenberg GD, Bird TD, Wijsman EM et al. Genetic linkage evidence for a familial Alzheimer's disease locus on chromosome 14. Science 1993; 258:668 671.

58. Sherrington R, Rogaev EI, Liang Y et al. Cloning of a gene bearing missense mutations in early-onset familial Alzheimer's disease. Nature 1995; 375:754-760.

59. Levy-Lahad E, Wijsman EM, Nemens E et al. A familial Alzheimer's disease locus on chromosome I. Science 1995a; 269:970-973.

60. Levy-Lahad E, Wasco W, Poorkaj P et al. Candidate gene for the chromosome 1 familial Alzheimer's disease locus. Science 1995b; 269:973-977.

61. Rogaev EI, Sherrington R, Rogaeva EA et al. Familial Alzheimer's disease in kindreds with missense mutations in a gene on chromosome 1 related to the Alzheimer's disease type 3 gene. Nature1995; 376:775-778.

62. Li J, Ma J, Potter H. Identification and expression analysis of a potential familial Alzheimer's disease gene on chromosome 1 related to AD3. Proc Natl Acad Sci USA 1995; 92:12180-12184.

63. Migliore L, Botto N, Scarpato R et al. Preferential occurrence of chromosome 21 malsegregation in peripheral blood lymphocytes of Alzheimer's disease patients. Cytogenet Cell Genet 1999; 87:41-46.

64. Yang Y, Geldmacher DS, Herrup K. DNA replication precedes neuronal cell death in Alzheimer's disease. J Neurosci 2001; 21:2661-2668

65. Ford JH. Spindle microtubular dysfunction in mothers of Down syndrome children. Hum Genet 1984; 68:295-298.

66. Staessen C, Maes AM, Kirsch-Volders M et al. Is there a predisposition for meiotic non-disjunction that may be detected by mitotic hyperploidy? Clinical Genetics 1983; 24:184-190.

67. Suzuki T, Oishi M, Marshak DR et al. Cell cycle-dependent regulation of the phosphorylation and metabolism of the Alzheimer amyloid precursor protein. EMBO J 1994; 13:1114-1122.

68. Pope WB, Lambert MP, Leypold B et al. Microtubule-associated protein tau is hyperphosphorylated during mitosis in the human neuroblastoma cell line SH-SY5Y. Exper Neurol 1994; 126:185-194.

69. Preuss U, Döring F, Illenberger S et al. Cell cycle-dependent phosphorylation and microtubule binding of tau protein stably transfected in Chinese hamster ovary cells. Mol Biol Cell 1995; 6:1397-1410.

70. Arendt T, Holzer M, Grossmann A et al. Increased expression and subcellular translocation of the mitogen activated protein kinase and mitogen-activated protein kinase in Alzheimer's disease. Neuroscience 1995; 68:5-18.

71. Arendt T, Rodel L, Gartner U et al. Expression of the cyclin-dependent kinase inhibitor p16 in Alzheimer's disease. NeuroReport 1996; 7:3047-3049.

72. Smith TW, Lippa CF. TI Ki-67 immunoreactivity in Alzheimer's disease and other neurodegenerative disorders. Journal of Neuropathology & Experimental Neurology 1995; 54:297-303.

73. Vincent I, Rosado M, Davies P. Mitotic mechanisms in Alzheimer's disease? J Cell Biol 1996; 132:413-425.

74. Vincent I, Jicha G, Rosado M et al. Aberrant expression of mitotic Cdc2/cyclin B1 kinase in degenerating neurons of Alzheimer's disease brain. J Neurosci 1997; 17:3588-3598.

75. Kondratick CM, Vandre DD. Alzheimer's disease neurofibrillary tangles contain mitosis-specific phosphoepitopes. J Neurochem 1996; 67:2405-2416.

76. Nagy Z, Esiri MM, Cato AM et al. Cell cycle markers in the hippocampus in Alzheimer's disease. Acta Neuropathologica 1997; 94:6-15.

77. McShea A, Harris PLR, Webster KR et al. Abnormal expression of the cell cycle regulators P16 and CDK4 in Alzheimer's disease. Amer J Pathol 1997; 150:1933-1939.

78. Li J, Xu M, Zhou H et al. Alzheimer presenilins in the nuclear membrane, interphase kinetochores, and centrosomes suggest a role in chromosome segregation. Cell 1997; 90:917-927.

79. McShea A, Wahl AF, Smith MA. Re-entry into the cell cycle: A mechanism for neurodegeneration in Alzheimer disease. Medical Hypotheses 1998; in press.

80. Schellenberg GD, Bird TD, Wijsman EM et al. Genetic linkage evidence for a familial Alzheimer's disease locus on chromosome 14. Science 1992; 258:668-71.

81. Kovacs DM, Fausett HJ, Page KJ et al. Alzheimer-associated presenilins 1 and 2: Neuronal expression in brain and localization to intracellular membranes in mammalian cells. Nature Med 1996; 2:224-229.

82. Walter J, Capell A, Grünberg J et al. The Alzheimer's disease-associated presenilins are differentially phosphorylated proteins located predominantly within the endoplasmic reticulum. Molec Med 1996. 2:673-91

83. Strooper B De, Beullens M, Contreras B et al. Phosphorylation, subcellular localization, and membrane orientation of the Alzheimer's disease-associated presenilins. J Biol Chem 1997; 272:3590-3598.

84. Moroi Y, Hartman AL, Nakane PK et al. Distribution of kinetochore (centromere) antigen in mammalian cell nuclei. J Cell Biol 1981; 90:254-259.

85. Brenner S, Pepper D, Berns MW et al. Kinetochore structure, duplication, and distribution in mammalian cells: analysis by human autoantibodies from scleroderma patients. J Cell Biol 1981; 91:95-102.

86. Weimer R, Haaf T, Kruger J et al. Characterization of centromere arrangements and test for random distribution in G0, G1, S, G2, G1, and early S' phase in human lymphocytes. Hum Genet 1992; 88:673-682.

87. Annaert WG, Levesque L, Craessaerts K et al. Presenilin 1 controls gamma-secretase processing of amyloid precursor protein in pre-golgi compartments of hippocampal neurons. J Cell Biol 1999; 147:277-294.

88. Honda T, Nihonmatsu N, Yasutake K et al. Familial Alzheimer's disease-associated mutations block translocation of full-length presenilin 1 to the nuclear envelope. Neurosci Res 2000; 37:101-111

89. Jeong SJ, Kim HS, Chang KA et al. Subcellular localization of presenilins during mouse preimplantationdevelopment. FASEB J 2000; 14(14):2171-2176.

90. Johnsingh AA, Johnston JM, Merz G et al. Altered binding of mutated presenilin with cytoskeleton-interacting proteins. FEBS Lett 2000; 465(1):53-58.

91. Pigino G, Pelsman A, Mori H et al. Presenilin-1 mutations reduce cytoskeletal association, deregulate neurite growth, and potentiate neuronal dystrophy and tau phosphorylation. J Neurosci 2001; 21:834-842

92. Janicki SM, Monteiro MJ. Presenilin overexpression arrests cells in the G1 phase of the cell cycle: arrest potentiated by the Alzheimer's disease PS2 (N141I) Mutant. Am J Pathol 1999; 155:135-144.

93. Vito P, Lacana E, D'Adamio L. Interfering with apoptosis: Ca(2+)-binding protein ALG-2 and Alzheimer's disease gene ALG-3. Science 1996; 271:521-165.

94. Vito P, Wolozin B, Ganjei JK et al. Requirement of the familial Alzheimers disease gene PS2 for apoptosis. Opposing effect of ALG-3. J Biol Chem 1996; 271: 31025-31028.

95. Wolozin B, Iwasaki K, Vito P et al. Participation of presenilin 2 in apoptosis: Enhanced basal activity conferred by an Alzheimer mutation. Science 1996; 274:1710-1713.

96. Wragg M, Hutton M, Talbot C. Genetic association between intronic polymorphism in presenilin-1 gene and late-onset Alzheimer's disease. Alzheimer's Disease Collaborative Group. Lancet 1996; 347(9000):509-512.

97. Higuchi S, Muramatsu T, Matsushita S et al. Presenilin-1 polymorphism and Alzheimer's disease. Lancet 1996; 347:1186.

98. Isoe K, Urakami K, Ji Y et al. Presenilin-1 polymorphism in patients with Alzheimer's disease, vascular dementia and alcohol-associated dementia in Japanese population. Acta Neurol Scand 1996; 94:326-328.

99. Kehoe P, Williams J, Holmans P et al. Association between a PS-1 intronic polymorphism and late onset Alzheimer's disease. NeuroReport 1996a; 7:2155-2158.

100. Kehoe P, Williams J, Lovestone S et al. Presenilin-1 polymorphism and Alzheimer's disease. Lancet 1996b; 347:1185.

101. Brookes AJ, Howell WM, Woodburn K et al. Presenilin-I, presenilin-II, and VLDL-R associations in early onset Alzheimer's disease. Lancet 1997; 350:336-337.

102. Ezquerra M, Blesa R, Tolosa E et al. The genotype 2/2 of the presenilin-1 polymorphism is decreased in Spanish early-onset Alzheimer's disease. Neurosci Lett 1997; 227:201-204.

103. Nishiwaki Y, Kamino K, Yoshiiwa A et al. T/G polymorphism at intron 9 of presenilin 1 gene is associated with, but not responsible for sporadic late-onset Alzheimer's disease in Japanese population. Neurosci Lett 1997; 227:123-126.

104. Tilley L, Morgan K, Grainger J et al. Evaluation of polymorphisms in the presenilin-1 gene and the butyrylcholinesterase gene as risk factors in sporadic Alzheimer's disease. Eur J Hum Genet 1999; 7:659-663.

105. Perez-Tur J, Froelich S, Prihar G et al. A mutation in Alzheimer's disease destroying a splice acceptor site in the presenilin-1 gene. Neuroreport 1995; 7.297-301.

106. Scott WK, Growdon JH, Roses AD et al. Presenilin-1 polymorphism and Alzheimer's disease. Lancet 1996; 347:1186-1187.

107. Scott WK, Yamaoka LH, Locke PA et al. No association or linkage between an intronic polymorphism of presenilin-1 and sporadic or late-onset familial Alzheimer disease. Genet Epidemiol 1997; 14.307-315.

108. Cai X, Stanton J, Fallin D et al. No association between the intronic presenilin-1 polymorphism and Alzheimer's disease in clinic and population-based samples. Am J Med Genet 1997; 74:202-203.

109. Lendon CL, Myers A, Cumming A et al. A polymorphism in the presenilin 1 gene does not modify risk for Alzheimer's disease in a cohort with sporadic early onset. Neurosci Lett 1997; 228:212-214.

110. Singleton AB, Lamb H, Leake A et al. No association between an intronic polymorphism in the presenilin-1 gene and Alzheimer's disease. Neurosci Lett 1997; 234:19-22

111. Sorbi S, Nacmias B, Tedde A et al. Presenilin-1 gene intronic polymorphism in sporadic and familial Alzheimer's disease. Neurosci Lett 1997; 222:132-134.

112. Tysoe C, Galinsky D, Robinson D et al. Analysis of α-1 antichymotrypsin, presenilin-1, angiotensin-converting enzyme, and methylenetetrahydrofolate reductase loci as candidates for dementia. Am J Med Genet 1997; 74:207-212.

113. Tysoe C, Whittaker J, Cairns NJ et al. Presenilin-1 intron 8 polymorphism is not associated with autopsy-confirmed late-onset Alzheimer's disease. Neurosci Lett 1997b; 222:68-69.

114. Wu X, Jiang S, Lin S et al. No association between the intronic presenilin 1 polymorphism and Alzheimer's disease in the Chinese population. Am J Med Genet 1999; 88:1-3

115. Petersen MB, Karadima G, Samaritaki M et al. Association between presenilin-1 polymorphism and maternal meiosis II errors in Down syndrome. Am J Med Genet 2000; 93:366-72.

116. Courvalin JC, Segil N, Blobel G et al. The lamin B receptor of the inner nuclear membrane undergoes mitosis-specific phosphorylation and is a substrate for $p34^{cdc2}$-type protein kinase. J Biol Chem 1992; 267:19035-19038.

117. Winey M, Hoyt MA, Chan C et al. NDC1: A nuclear periphery component required for yeasts spindle pole body duplication. J Cell Biol 1993; 122:743-751.

118. Ye Q, Worman HJ. Primary structure analysis and lamin B and DNA binding of human LBR, an integral protein of the nuclear envelope inner membrane. J Biol Chem 1994; 269:11306-11311.

119. Thinakaran G, Borchelt DR, Lee MK et al. Endoproteolysis of presenilin 1 and accumulation of processed derivatives in vivo. Neuron 1996; 17:181-190.

120. Mercken M, Takahashi H, Honda T et al. Characterization of human presenilin 1 using N-terminal specific monoclonal antibodies: evidence that Alzheimer mutations affect proteolytic processing. FEBS Letters 1996; 389:297-303.

121. Shen J, Bronson RT, Chen DF et al. Skeletal and CNS defects in Presenilin-1-deficient mice. Cell 1997; 89:629-639.

122. Duff K, Eckman C, Zher C et al. Increased amyloid-β42(43) in brains of mice expressing mutant presenilin 1. Nature 1996; 383:710-713.

123. Donehower LA, Godley LA, Aldaz CM et al. Deficiency of p53 accelerates mammary tumorigenesis in Wnt-1 transgenic mice and promotes chromosomal instability. Dev 1995; 9:882-895.

124. Fukasawa K, Choi T, Kuriyama R et al. Abnormal centrosome amplification in the absence of p53. Science 1996; 271:1744-1747.

125. Donehower LA. Genetic instability in animal tumorigenesis models. Cancer Surv 1997; 29:329-352.

126. Boyle JM, Mitchell ELD, Graves MJ et al. Chromosome instability is a predominant trait of fibroblasts from Li-Fraumeni families. Br J Cancer 1998; 77:2181-2192.

127. Roperch JP, Alvaro V, Prieur S et al. Inhibition of presenilin 1 expression is promoted by p53 and p21WAF-1 and results in apoptosis and tumor suppression. Nat Med 1998; 14:835-838.

128. Altman J, Das GD. Autoradiographic and histologic evidence of postnatal neurogenesis in rats. J Comp Neurol 1965; 124:319-335.

129. Kaplan MS, Hinds JW. Neurogenesis in the adult rat: electron microscopic analysis of light radioautographs. Science 1977; 197:1092-1094.

130. Bayer SA, Yackel JW, Puri PS. Neurons in the rat dentate gyrus granular layer substantially increase during juvenile and adult life. Science 1982; 216:890-892.

131. Cameron HA, Woolley CS, McEwen BS et al. Differentiation of newly born neurons and glia in the dentate gyrus of the adult rat. Neuroscience 1993; 56:337-344.

132. Kaplan MS, Bell DH. Mitotic neuroblasts in the 9-day-old and 11-month-old rodent hippocampus. J Neurosci 1984; 4:1429-1441.

133. Gage FH, Ray J, Fisher LJ. Isolation, characterization, and use of stem cells from the CNS. Annu Rev Neurosci 1995; 18:159-192.

134. Kuhn GH, Dickinson-Anson H, Gage FH. Neurogenesis in the dentate gyrus of the adult rat: age-related decrease of neuronal progenitor proliferation. J Neurosci 1996; 16(6):2027-2033.

135. Gould E, Tanapat P, McEwen BS et al. Neurogenesis in the dentate gyrus of adult primates can be suppressed by stress. Proc Natl Acad Sci USA 1998; 95:3168-3171.

136. Gould E, Reeves AJ, Graziano MS et al. Neurogenesis in the neocortex of adult primates. Science 1999a; 286:548-552.

137. Gould E, Reeves AJ, Fallah M et al. Hippocampal neurogenesis in adult Old World primates. Proc Natl Acad Sci USA 1999b; 96:5263-5267.

138. Eriksson PS, Perfilieva E, Bjork-Eriksson T et al. Neurogenesis in the adult human hippocampus. Nature Med 1998; 4:1313-1317.

139. LeBlank A. Increased production of 4KDa amyloid beta peptide in serum-deprived human primary neuron cultures: possible involvement of apoptosis. J Neurosci 1995; 15:7837-7846.

140. Galli C, Piccini A, Ciotti MT et al. Increased amyloidogenic secretion in cerebellar granule cells undergoing apoptosis. Proc Natl Acad Sci USA 1998; 95:1247-1252.

141. Rakic, P. Limits of neurogenesis in primates. Science 1985; 227:1054-1056

142. McGeer PL, Itagaki S, Boyes E et al. Reactive microglia in patients with senile dementia of the Alzheimer type are positive for the histocompatibility glycoprotein HLA-DR. Neurosci Lett 1987; 79:195-200.

143. Abraham CR, Selkoe DJ, Potter H. Immunochemical identification of the serine protease inhibitor α_1-antichymotrypsin in the brain amyloid deposits of Alzheimer's disease. Cell 1988; 52:487-501.

144. Abraham CR, Potter H. Alzheimer's disease: Recent advances in understanding the brain amyloid deposits. Biotechnology 1998; 7:147-153.

145. Potter H. The involvement of astrocytes and an acute phase response in the amyloid deposition of Alzheimer's disease. In: Yu ACH, Hertz LMD, eds. Neuronal-astrocytic interactions: Implication for normal and pathological CNS function. Prog Brain Res 1992; 94:447-458.

146. McGeer PL, Kawamata T, Walker DG et al. Microglia in degenerative neurological disease. Glia 1993; 7:84-92.

147. Eikelenboom P, Zhan SS, van Gool WA et al. Inflammatory mechanisms in Alzheimer's disease. TIPS 1994; 15:447-450.

148. Das S, Potter H. Expression of the Alzheimer amyloid-promoting factor antichymotrypsin is induced in human astrocytes by IL-1. Neuron 1995; 14:447-456.

149. Ma J, Brewer Jr HB, Potter H. Alzheimer Aβ neurotoxicity: promotion by antichymotrypsin, ApoE4, inhibition by Aβ-related peptides. Neurobiol of Aging 1996; 17:773-780.

150. Griffin WST, Stanley LC, Ling C et al. Brain interleukin-1 and S100 immunoreactivity elevated in Down's syndrome and Alzheimer's disease. Proc Natl Acad Sci USA 1989; 86:7611-7615.

151. Sheng JG, Mrak RE, Griffin WS. Microglial interleukin-1 alpha expression in brain regions in Alzheimer's disease: correlation with neuritic plaque distribution. Neuropathol Appl Neurobiol 1995; 21:290-301.

152. Sanan DA et al. Apolipoprotein E associates with beta amyloid peptide of Alzheimer's disease to form novel monofibrils. Isoform apoE4 associates more efficiently than apoE3. J Clin Invest 1994; 94:860-869.

153. Castano EM, Frangione B. Biology of disease: Human amyloidosis, Alzheimer disease and related disorders. Lab Invest 1988; 58:122-132.

154. Selkoe DJ. Molecular pathology of amyloidogenic proteins and the role of vascular amyloidosis in Alzheimer's disease. Neurobiol Aging 1989b; 10:387-3395.

155. Busciglio J, Yankner BA. Apoptosis and increased generation of reactive oxygen species in Down's syndrome neurons in vitro. Nature 1997; 378:776-779.

156. Teller JK, Russo C, DeBusk LM et al. Presence of soluble amyloid β-peptide precedes amyloid plaque formation in Down's syndrome. Nature Medicine 1996; 2:93-95.

157. Deng G, Pike CJ, Cotman CW. Alzheimer-associated presenilin-2 confers increased sensitivity to apoptosis in PC12 cells. FEBS Letters 1996; 397:50-54.

158. Janicki S, Monteiro MJ. Increased apoptosis arising from increased expression of the Alzheimer's disease-associated presenilin-2 mutation (N141I). J Cell Biol 1997; 139:485-495.

159. Martins RN, Turner BA, Carroll RT et al. High levels of amyloid-β protein from S182 (Glu246) familial Alzheimer's cells. NeuroReport 1995; 7:217-220.

160. Scheuner D, Eckman C, Jensen M et al. Secreted amyloid β-protein similar to that in the senile plaques of Alzheimer's disease is increased in vivo by the presenilin 1 and 2 and APP mutations linked to familial Alzheimer's disease. Nature Med 1996; 2:864-870.

161. Borchelt DR, Thinakaran G, Eckman CB et al. Familial Alzheimer's disease-linked presenilin 1 variants elevate Aβ1-42/1-40 ratio in vitro and in vivo. Neuron 1996; 17:1005-1013.

162. Citron M, Westaway D, Xia W et al. Mutant presenilins of Alzheimer's disease increase production of 42-residue amyloid β-protein in both transfected cells and transgenic mice. Nature Med 1997; 3:67-72.

163. Suzuki N, Cheung TT, Cai XD et al. An increased percentage of long amyloid β protein secreted by familial amyloid β protein precursor (βAPP$_{717}$) mutants. Science 1994b; 264:1336-1340.

164. Yamatsuji T, Matsui T, Okamoto T et al. G protein-mediated neuronal DNA fragmentation induced by familial Alzheimer's disease-associated mutants of APP. Science 1996; 272:1349-1352.

165. Yamatsuji T, Okamoto T, Takeda S et al. Expression of V642 APP mutant causes cellular apoptosis as Alzheimer trait-linked phenotype. EMBO J 1996; 15:498-509.

166. Zhao B, Chrest FJ, Horton WE, Jr. et al. Expression of mutant amyloid precursor proteins induces apoptosis in PC12 cells. J Neurosci Res 1997; 47:253-263.

167. Gandy S, Czernik AJ, Greengard P. Phosphorylation of Alzheimer disease amyloid precursor protein peptide by protein kinase C and Ca^{2+}/calmodulin-dependent protein kinase II. Proc Natl Acad Sci USA 1988; 85:6218-6221.

168. Buxbaum JD, Gandy SE, Cicchetti P et al. Processing of Alzheimer b/A4 amyloid precursor protein: modulation by agents that regulate protein phosphorylation. Proc Natl Acad Sci USA 1990; 87:6003-6006.

169. Demaerschlack I, Delvaux A, Octave JN. Activation of protein kinase C increases the extracellular release of the transmembrane amyloid protein precursor of Alzheimer's disease. Biochimica et Biophysica Acta 1993; 1181:214-218.

170. da Cruz e Silva EF, da Cruz e Silva OAB, Zaia CTBV et al. Inhibition of protein phosphatase 1 stimulates secretion of Alzheimer amyloid precursor protein. Mol Med 1995; 1:535-541.

171. Xu H, Sweeney D, Greengard P et al. Metabolism of Alzheimer b-amyloid precursor protein: regulation by protein kinase A in intact cells and in a cell-free system. Proc Natl Acad Sci USA 1996; 93:4081-4084.

172. Autilio-Gambetti L, Morandi A, Tabaton M et al. The amyloid percursor protein of Alzheimer disease is expressed as a 130 kDa polypeptide in various cultured cell types. FEBS Lett 1988; 241:94-98.

173. Dranovsky A, Oxberry W, Lyubsky S et al. APP is associated with mitotic centrosomes and a distinct cytoplasmic network which colocalizes with depolymerized tubulin. Neurobiol Aging 1996; 17:S196 (abs. 788).

174. Weidemann A, Paliga K, Durrwang U et al. Formation of stable complexes between two Alzheimer's disease gene products: presenilin-2 and beta-amyloid precursor protein. Nature Med 1997; 3:328-323.

175. Chen Y, McPhie DL, Hirschberg J et al. The amyloid precursor protein-binding protein APP-BP1 drives the cell cycle through the S-M checkpoint and causes apoptosis J Biol Chem 2000; 275:8929-8935.

176. von der Kammer H, Hanes J, Klaudiny J et al. A human amyloid precursor-like protein is highly homologous to a mouse sequence-specific DNA-binding protein. DNA Cell Biol 1994; 13:1137-1143.

177. Rassoulzadegan M, Yang Y, Cuzin F. APLP2, a member of the Alzheimer precursor protein family, is required for correct genomic segregation in dividing mouse cells. EMBO J 1998; 17:4647-4656.

178. Avramopoulos D, Mikkelsen M, Vassilopoulos D et al. Apolipoprotein E allele distribution in parents of Down's syndrome children. Lancet 1996; 347:862-865.

179. Strittmatter WJ, Roses D. Apolipoprotein E and Alzheimer disease. Proc Natl Acad Sci USA 1995; 92:4725-4727.

180. Mistry MJ, Clay MA, Kelly ME et al. Apolipoprotein E restricts interleukin-dependent T lymphocyte proliferation at the G1A/G1B boundary. Cell Immunol 1995; 160:14-23.

181. Uchida IA, Lee CPV, Byrnes EM. Chromosome aberrations in vitro by low doses of radiation: nondisjunction in lymphocytes of young adults. Am J Hum Genet 1975; 27:419-429.

182. Martyn CN, Osmond C, Edwardson JA et al. Geographical relation between Alzheimer's disease and aluminium in drinking water. Lancet 1989; January:59-62.

183. Palekar LD, Eyre JF, Most BM et al. Metaphase and anaphase analysis of V79 cells exposed to erionite, UICC chrysotile, and UICC crocidolite. Carcinogen 1987; 8:553-560.

184. Ganrot PO. Metabolism and possible health effects of aluminum. Environ Health Persp 1986; 65:363-441.

185. Holm C, Stearns T, Botstein D. DNA topoisomerase II must act at mitosis to prevent nondisjunction and chromosome breakage. Mol Cell Biol 1898; 9:159-168.

186. Cai M, Davis RW. Yeast centromere binding protein CBF1, of the helix-loop-helix protein family, is required for chromosome stability and methionine prototrophy. Cell 1990; 61:437-446.

187. Rockmill B, Fogel S. DISI: A yeast gene required for proper meiotic chromosome disjunction. Genetics 1988; 119:261-272.

188. Barinaga M. Missing Alzheimer's gene found. Science 269:917-918

CHAPTER 6

Alzheimer Disease:

A New Beginning or a Final Exit?

Mark E. Obrenovich, Arun K. Raina, Osamu Ogawa, Craig S. Atwood, Laura Morelli and Mark A. Smith

Abstract

Today, a new chapter is being written in the book of Alzheimer disease, one that is challenging the longstanding view that adult neurons are incapable of division, remain nonproliferative, and are terminally differentiated. Here, we review the provocative notion that, in Alzheimer disease, whole populations of nonstem cell neurons leave their quiescent state and re-enter into the cell cycle. However, such neuronal re-entry into the cell cycle is futile and ultimately leads to the neurodegeneration that typifies Alzheimer disease.

Introduction

It is perhaps ironic to discover that neurodegenerative diseases, such as Alzheimer disease (AD), where cell loss is a key feature, may provide clues to understanding the plasticity of the adult central nervous system. In AD, there is accumulating evidence that susceptible neuronal populations exhibit a de-differentiated phenotype likely representative of a reactivated cell cycle. This exit from a quiescent state is manifested in several ways, including:

1. The ectopic expression of cyclins along with their cognate cell cyclin-dependent kinases (CDKs) and their inhibitors (CDKIs);
2. Recruitment of mitogenic signal transduction pathway components; and
3. The increased transcriptional activation of a variety of mitosis-related proteins.

While the cause of this apparent neuronal re-entry into the cell cycle is not known, the consequences for these terminally differentiated cells are disastrous leading to oxidative stress, cytoskeletal abnormalities, mitochondrial dysfunction and, ultimately, neuronal death. In other words, the re-emergence into the cell cycle by neurons accounts for many of the cardinal features of the disease. In this review, we explore some of these mitotic alterations including the recruitment of mitogenic factors and oxidative stress. Further, we speculate on the nature and the source(s) of mitogenic factors, which underlie the pathological events observed in this dreaded disease.

Pathological Hallmarks of Alzheimer Disease

As an insidious and progressive neurodegenerative disease, AD affects up to 15% of individuals over the age of 65 and nearly half of all individuals aged 85 and above.[1] The disease is quickly becoming one of the most serious health problems in the U.S. and has a dehumanizing

Cell Cycle Mechanisms and Neuronal Cell Death, edited by Agata Copani
and Ferdinando Nicoletti. ©2005 Eurekah.com and Kluwer Academic / Plenum Publishers.

nature that involves destruction of higher-order brain function leading to dementia, disability and, finally, death. Two pathological lesions, namely the neurofibrillary tangle (NFT) and the senile plaque, are hallmarks of the disease and these neuritic plaques and NFT are largely associated with dementia. NFT, which contain a highly phosphorylated form of the microtubule associated protein tau, is the major intracellular pathology of AD, while senile plaques are extracellular and are primarily composed of amyloid-β.[2,3] The mechanisms involved in the formation of these lesions or neuronal death are largely unknown although recent findings indicate a key role for the aberrant re-entry of neurons into the cell cycle.

A Mitotic Phenotype Appears in Alzheimer Disease

A growing number of cell cycle-related proteins are found associated with the susceptible and vulnerable neurons of AD (Table 1) that, from their temporal and pathological distribution, are indicative of an early and fundamental role in the pathogenesis of AD. This cycling phenotype, rather than a phenotype of cells in a terminally differentiated state, has been reviewed elsewhere.[4] Nonetheless, it is important to note that cell growth changes ultimately occur through signal transducers that activate specific transcription factors and modulate cell cycle control proteins.[5] These proteins themselves are also regulated in a cell cycle-dependent manner[6,7] and are listed in Table 2. Perhaps of greatest import, however, as regards disease pathogenesis, all of the major genetic and protein elements dysregulated in AD, including tau, amyloid-β precursor protein (AβPP), presenilin1/2, and, possibly, apolipoprotein E (ApoE), are also altered during the cell cycle.[4]

Tau Phosphorylation

Since increased phosphorylation and decreased microtubule stability are coincident during progression through the cell cycle[8,9] and these cell cycle-related protein alterations are found in AD, it is not surprising that microtubular abnormalities and tau phosphorylation are associated with AD.[10] While the kinases responsible for tau phosphorylation in AD are not completely characterized, increased residue-specific phosphorylation of tau occurs in mitotically active neurons where phosphorylation is driven by CDKs.[11-15] Of note, in AD, CDKs, such as CDK2 and CDK5, as well as Cdc-kinases and MAP2 kinases, are increased in AD in a topographical manner that completely overlaps with phospho-tau[16,17] and also have been shown to hyperphosphorylate tau in in vitro assays.[18-22] In addition, we recently demonstrated that CDK7, an age-dependent CDK-activating kinase, also associates phospho-tau in AD and may be essential to all other mitotic alterations since CDK7 plays such a crucial role as an activator of all the major CDK/Cyclin substrates.[23] Finally, we have shown that cell cycle re-entry leads to tau phosphorylation in primary neurons (McShea and Smith, unpublished data).

Amyloid-β

The major protein component of senile plaques, amyloid-β, is derived from a larger precursor AβPP encoded on chromosome 21[24] and is upregulated secondary to mitogenic stimulation.[25] Further, AβPP metabolism is regulated by cell cycle-dependent changes[7] and has neurotrophic effects at low (nM) concentrations[26] consistent with its mitogenic activity in vitro.[27,28] Presumably, the effect of amyloid-β is mediated through mitogen activated protein kinase (MAPK),[29] and therefore may play a direct role in the induction or propagation of cell cycle-mediated events in AD. Therefore, amyloid-β, along with oxidative stress[30] and cell cycle re-entry, may have common etiologies. However, it is notable that, while amyloid-β-mediated cell death, at least in vitro, is dependent on the presence of various cell-cycle-related elements,[31] in vivo analysis of the basal nucleus of Meynert and the locus ceruleus, where amyloid-β is rarely seen, found little or no topographical relationship between amyloid-β and the ectopic expression of cell cycle markers in diseased brains.[23,32-38] Thus, amyloid-β may only become toxic in vivo when the neuronal cell cycle machinery is activated or when levels exceed the body's ability to regulate its turnover.

Presenilins

Mutations in the human presenilin genes 1 and 2 (PS-1/2) found on chromosomes 14 and 1, respectively, are linked to early onset AD.[1] The association of presenilins with centrosomes and centromers, and the link between PS-1/2 and Notch-based signaling through cadherin-based cell-cell adherence junctions,[39] indicates that PS-1/2 may play critical roles in cytoskeletal anchorage, cell division, chromosome segregation,[40] cell fate,[41-43] early embryonic development,[44] and tumorigenesis.[45] In this regard, we and others have shown that overexpression of PS-1/2 leads to cell arrest in the G_1 phase of the cell cycle, an effect that is potentiated by expression of the PS-2 (N141I) mutation.[46] Overexpression also yields a loss of calcium homeostasis, increased oxidative stress, and increased susceptibility to apoptotic death,[47] with AD-linked mutations of presenilins showing greater effect.[46,48] Further, PS-1 mutations destabilize beta-catenin and can potentiate neuronal apoptosis,[49] by reducing the capacity of neurons to induce endoplasmic reticulum chaperones.[50] Alternatively, induction of apoptotic systems via PS-1/2 and AβPP mutations, could also lead to the upregulation of CDKs since expression of Cyclin/CDKs, in addition to driving cell proliferation and growth control, are also implicated in neuronal death signaling and apoptosis (see Fig. 1).[31,51,52] Indeed, the differential activity of AD-linked PS-1 in the beta-catenin-signaling pathway indicates a key role for cadherins in the pathogenesis of AD. Therefore, one would also expect the subsequent induction of p27 (Ogawa et al, submitted) and inhibition of Cyclin E/CDK2, while increasing expression of p21[53] and consequently an inhibition of proliferation. In addition, a block from progression at the G1/S phase boundary, by PS (and possibly AβPP) mutations, would likely result in the accumulation of cell cycle control proteins as is seen in AD. Therefore, PS mutations confer a contracted time course to the underlying pathophysiology of AD.

AβPP, through the stimulation of Ras-dependent MAPK cascade in vivo, is correlated with highly phosphorylated tau.[54] The early p21Ras expression pathway is activated during the posttranslational modification of AβPP and tau phosphorylation, which precedes neurofibrillary degeneration and amyloid-β formation.[55] Additionally, the presence of p21, highly phosphorylated tau, Ki-67, and cell cycle-associated nuclear antigen protein (PCNA), may have a role in the production of abnormally phosphorylated tau which then leads to the formation of cytoskeletal derangements in susceptible neurons.[56] This strong link points to cell cycle reactivation and the upstream ectopic expression of cell cycle markers as a critical, and common, early event in AD pathogenesis.

G_0 Exit, G_1 Entry, and Mitogenic Drivers in Alzheimer Disease

Quiescence, cell division, and differentiation are states central to the regulation of growth and development. Increased growth stimuli, as in extrinsic mitotic pressure, activate key factors for G_0 exit and G_1 progression, including the complex-forming CDKs, i.e., CDKs 4, 5, 6, 7 (see Table 1), and their cognate activating cyclins, i.e., Cyclin D1, D3, E and B1 (see Table 1). These complexes are able to phospho-regulate a wide variety of relevant substrates.[57] Together, they orchestrate DNA replication, cytoskeletal re-organization, and cellular metabolism required for proliferation, development, and cell cycle progression. While it has been argued that a number of the cell-cycle related phenomena found in AD can also occur as sequelae to other processes, such as apoptosis, trophic-deprivation, and DNA repair (see Table 2),[52,58-62] we propose that the re-emergence of, or sensitivity to, extrinsic signals initiates an attempt to re-enter into the cell division cycle, with progression being limited by the degree of mitotic competence of the adult neuron.

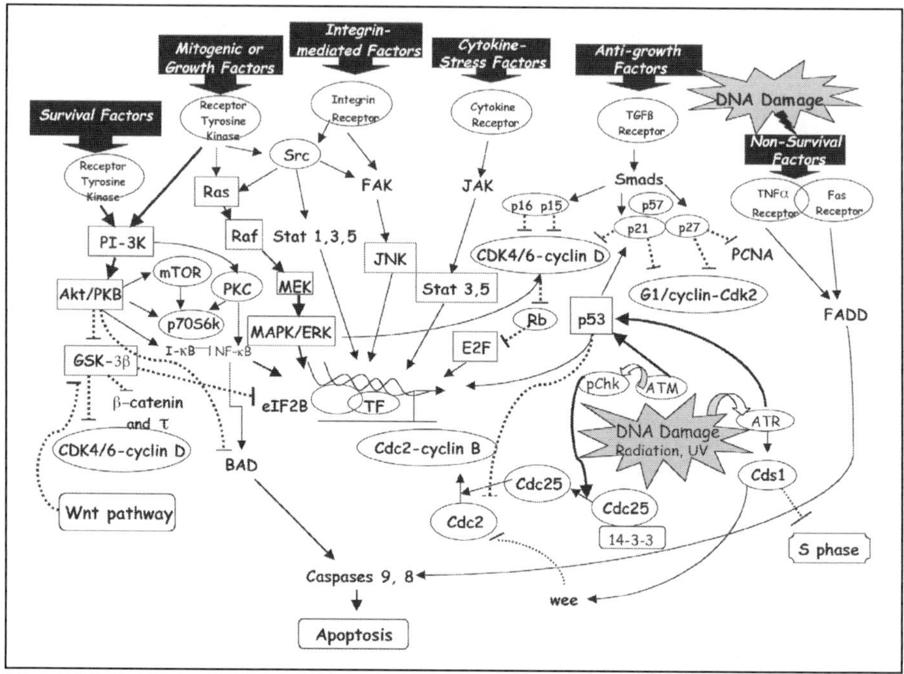

Figure 1. The complex relationship between cell survival and cell death pathways.

The identity of the signal(s) that lead the neuron to attempt exit from a quiescent state and re-enter the cell cycle remains yet to be determined. However, a number of growth factors and mitogens are elevated in the AD brain and may drive cell cycle re-entry. Re-sensitization to these exogenous or surface-derived signals can lead to the activation of the mitotic engine and drive cell proliferation as seen in AD. Candidate growth factors, elevated in the AD brain include, but are not limited to, neurotrophic factors, nerve growth factor (NGF), transforming growth factor beta-1 (TGF-β),[63] platelet-derived growth factor (PDGF), epidermal growth factor (EGF), and basic fibroblast growth factor (bFGF).[64-67] Additionally, insulin-like growth factor-1, which has been shown to mediate transient site-selective increases in tau phosphorylation in primary cortical neurons,[68] is involved in axonal growth and development and can mediate the cytoskeletal reorganization that occurs during neurite outgrowth and, perhaps, in aberrant neuronal sprouting.[69]

Apoptotic Avoidance in Alzheimer Disease

Apoptotic avoidance by itself can be viewed as both sufficient and necessary for transformative processes. Therefore, we made a systematic study of the caspase cascade proteins in AD by evaluating the presence and/or absence of central initiator (caspases 8 and 9) and the executioner (caspases 3, 6 and 7) proteins of apoptosis.[165] Our study revealed that although upstream initiator caspases were present in association with the pathological lesions in all cases of AD, downstream executioner caspases, including 3 and 7, that signal the onset of the execution phase of apoptosis, remained at control levels in vulnerable populations indicating an absence of effective distal propagation of the caspase-mediated apoptotic signal(s). This lack of downstream amplification of signaling via the caspase pathway may well account for the lack of an apoptotic phenotype but the development of a mitotic phenotype in AD. Notably, expression

Table 1. Cell cycle-specific markers found associated with Alzheimer disease

Marker	Role	Association with Alzheimer Disease
Cyclin A	S to G2/M	6,70
Cyclin B	G2/M	21,22,35,71
Cyclin C	No Known Role	
Cyclin D (D3)	G0/G1/ lateG1/S	35,71-73
Cyclin E	G1 to G1/S	21,22,74
p34cdc2/ cdk 1	Late G2/M	6,7,16,35,70,71,75
Cdk4/Cdk6	G1/ G1/S	31,32,35,75,76
Cdk5/p25/p35	G2 D1, D3 G1 Cyclins	17,76-83
Nclk cdc2-like kinase	Cyclin A kinase	17,20,76,84
Cdk7/MPM2	CDK activated kinase	23,72,73
Cdc42/rac	GTPase/cell division	36
p21ras	G protein/MAPK	46,54,55,70,85
MRG 15	M phase regulator	86
Ki-67	LateG1,S,G2,M	21,22,56,87
p105/pRb	G2/M TF	31,56,87
pCNA	Non cell-cycle specific	35
p107/pRb	Cdk2/4/6, check pt	31,48 (negative association)
c-myc	S to G2 checkpoint	46 (negative association)
p53/MDM2	Repressor complex	88,89
ATM	Check-point	46
Raf/Raf-1	Check point kinase	90
p16INK4a p18p15p19	CyclinD/cdk4/6 inhibitors of M phase	19,32,70,72,73,91
p27/Kip1	Cyclin D and E /cdk7 inhibitor	46 (negative association),72,73
WAF-1/ p21/Cip1	Multi-Cyclin /cdk-inihibitor (G1 and S)	92
Polo-like kinase	G2/M check point	93,94

of cyclin/CDKs, in addition to driving cell proliferation and growth control, has dual conserved roles as they are also implicated in apoptotic signaling.[31,51,52]

Redox Imbalance and Cell Cycle Re-Entry

Energy is an obligate requirement for dividing cells. Therefore, before mitosis, there is division and redistribution of cellular organelles such that during late S, G2 and mitotic phases, mitochondrial proliferation is most evident.[166] Notably, in AD, increases in the number of mitochondria are found in the same neurons that also exhibit cell cycle related abnormalities and undergo subsequent oxidative damage and cell death.[167] While in a normally mitotic cell, mitochondrial replication is imperative for providing the energy needed for cell division, in AD where neuronal cell cycle is interrupted or dysfunctional, we suspect that neurons incur a "phase stasis" with excessive mitochondria. Such "excess" mitochondria are then potent sources of free radicals and cause homeostatic and redox imbalances, especially in those redox reactions involving calcium metabolism.[168] Thus, cell cycle dysfunction, when mitochondrial mass is highest, poses an elevated, and possibly chronic, oxidative assault upon the cell, far beyond the blunting capacity of endogenous antioxidants.

Importantly, imbalances in redox homeostasis are also played out via numerous signal transduction cascades, which are also intimately linked to cell cycle control. Indeed, activation of p38 MAPK and ERK links tau phosphorylation, oxidative stress, and cell cycle-related events

Table 2. Cell cycle-associated proteins found in Alzheimer disease

Marker	Role	Association with Alzheimer Disease
PP2A or PP2B	Phosphatase (Cdk5, cdc2)	76,95-98
PP-1		81,95
Cdc25 Cdc25A	Phosphatase G2/M	99,100
PKC/ Wnt path	Translation control	101-109
PKA	Kinase	110,111
PKN	Kinase	112
PI3K	Kinase	113-116
AKT/PKB/RAC	Kinase	112,116-119
TGFBeta/ TAK	Kinase	120,121
p44/p42 MAPK (ERK1/2)	MAP kinase	16,38,54,70,122-135
CamK	Kinase Ca /Calmodulin regulated	136
p38 MAPK	Kinase	37,133,134,137-139
JNK/ (SAPK-2/3) - alpha gamma	Kinase (stress activated)	38,133,134,140
MEK	MAPK Kinase	70,126
GSK-3 and beta Catenin	Proline dependent protein kinase (PDPK)	17,49,76,77,80,81,109,113, 118,119, 122,133,137,141-151
P120/E-cadherin	Adhesion complex	152
c-fos	TF / regulator	153
14-3-3/14-3-3zeta	Adaptor protein	154,155
c-jun/p39, AP-1	TF component	101,153,156-159
Fyn	Transcription factor	160-162
p53	TF / DNA damage	21,22,163
Rho	G-protein	112,164
Rap Rab	G-protein	90
Sos-1	Guanine nucleotide exchange factor	33
Grb-2	Adaptor	33

in AD.[37,38,128,169] MEK, ERK1/2, cyclins, cyclin-dependent kinases and their inhibitors, i.e., p16INK4a family, and p21Ras are elevated early in AD and co-localize in pyramidal neurons with NFT.[70] Neuronal ERK is increased in AD, and phosphorylation, as well as phosphorylation of p38 and CREB, by nerve growth factor or epidermal growth factor, is differentially modulated by oxidative and other stresses.[170] In support of this notion, compromised mitochondrial function was found to lead to increased cytosolic calcium and to the activation of MAPKs (ERK1/2).[127] Likewise, activated forms of ERK are found decreased in cells overexpressing heme oxygenase-1 (HO-1), indicating that tau and HO-1 both serve overlapping protective roles in regulating oxidative stress.[132] Importantly, there is abundant evidence that oxidative stress and free radical damage plays an essential role in the pathogenesis of AD.[38,171,172] Therefore, it is notable that free radicals, free-radical generators, and antioxidants also act as crucial control parameters of the cell cycle.[173,174]

Finally, there is abundant support for the notion that imperfect clearance of proteins, damaged or modified by oxidation processes, contributes to cell death by interfering with essential cell functions.[175] Impairments in the ubiquitin-dependent protein degradation system, which is aimed at clearing and preventing the progressive accumulation of misfolded or aggregated and ubiquitinated proteins, is a cytopathological feature in many neurodegenerative

Figure 2. "Cycling towards dementia" involves cell cycle dysregulation in AD.

disorders, including AD.[176,177] In support of this notion, accumulation of phosphorylated neurofilaments and the increase in apoptosis-specific protein and phosphorylated c-Jun is induced by proteasome inhibitors.[178] This speaks to the importance of cellular context in this process, and the fact that the ubiquitin-proteasome pathway plays an important role in the regulation of critical cellular processes, which include the cell cycle, cytoskeletal organization, and gene transcription, e.g., c-fos, p53, p21 and p27. Such proteolysis is known to drive the cell cycle by regulating the oscillations in activity of CDKs and perturbations in this process also likely contribute to the dysregulated cell cycle seen in AD.[179] In this regard, any event that would also upset the balance between the signal transduction pathways for survival, or those for growth, as well as those for death or differentiation, would likely shift this delicate balance. Ultimately, this shift would determine the fate of select neuronal cells and population subsets by largely favoring one set of pathways over another. The net result of this cross table would impact survival or death to the cell and perhaps offer an explanation for the protracted time course, which is seen in most neurodegenerative diseases (see Fig. 1).

Other damaging factors like hyperglycemia, reducing sugars and the presence of reactive oxygen and nitrogen species can have a direct role in mediating protein crosslinking and, thus, the accumulation of undigested material in AD.[171,180,181] In support of this notion, caloric restriction has been shown to selectively modulate the age-associated induction of genes encoding proteins involved in inflammatory and stress responses.[182]

Conclusions

Cycling toward dementia requires an imbalance and despite their supposedly quiescent status, vulnerable neurons in AD display a cell cycle phenotype, albeit an aberrant one (see Fig. 2). Further, it is becoming increasingly apparent that an altered and protracted cell cycle stasis exists in susceptible neurons in AD. In fact, these "abnormalities" may be a partial response to the genotoxic stress and metabolic imbalance common in degenerating neurons. Therefore, any re-emergence of sensitivity to extrinsic signals, i.e., neurotrophic factors, may initiate an attempt to re-enter into the cell division cycle, with progression being limited by the degree of mitotic competence. Successful dysregulation of the cell cycle, coupled with a multilevel apoptotic avoidance system, fulfills both the sufficient and necessary criteria for the initiation of an oncogenic transformation and therefore, early in the course of AD, neurons likely face the recruitment of similar mechanisms,[4] i.e., AD is analogous to cancer. Unfortunate as it may be to our higher-order structures, this opera of cell cycle appears to be unsustainable in neurons and eventually leads to stasis in a specific phase of the cell division cycle, cellular dysfunction and in the end-death.

References

1. Smith MA. Alzheimer disease. In: Bradley RJ, Harris RA, eds. International Review of Neurobiology, Vol. 42, San Diego: Academic Press, Inc., 1998:1-54.
2. Iqbal K, Zaidi T, Thompson CH et al. Alzheimer paired helical filaments: bulk isolation, solubility, and protein composition. Acta Neuropathol 1984; 62:167-177.
3. Grundke-Iqbal I, Iqbal K, Tung Y-C et al. Abnormal phosphorylation of the microtubule-associated protein tau (τ) in Alzheimer cytoskeletal pathology. Proc Natl Acad Sci USA 1986; 83:4913-4917.
4. Raina AK, Zhu X, Rottkamp CA et al. Cyclin' toward dementia: cell cycle abnormalities and abortive oncogenesis in Alzheimer disease. J Neurosci Res 2000; 61:128-133.
5. Gutkind SJ, ed. Signaling Networks and Cell Cycle Control: The Molecular Basis of Cancer and Other Diseases. Totowa, New Jersey: Humana Press, 2000.
6. Drewes G, Lichtenberg-Kraag B, Doring F et al. Mitogen activated protein (MAP) kinase transforms tau protein into an Alzheimer-like state. EMBO J 1992; 11:2131-2138.
7. Suzuki T, Oishi M, Marshak DR et al. Cell cycle-dependent regulation of the phosphorylation and metabolism of the Alzheimer amyloid precursor protein. EMBO J 1994; 13:1114-1122.
8. Lindwall G, Cole RD. Phosphorylation affects the ability of tau protein to promote microtubule assembly. J Biol Chem 1984; 259:5301-5305.
9. Alonso AC, Grundke-Iqbal I, Iqbal K. Alzheimer's disease hyperphosphorylated tau sequesters normal tau into tangles of filaments and disassembles microtubules. Nat Med 1996; 2:783-787.
10. Terry RD, Gonatas NK, Weiss M. Ultrastructural studies in Alzheimer's presenile dementia. Am J Pathol 1964; 44:269-297.
11. Brion JP, Passarier H, Nunez J et al. Immunologic determinants of tau protein are present in neurofibrillary tangles of Alzheimer's disease. Arch Biol 1985; 95:229-235.
12. Brion JP, Octave JN, Couck AM. Distribution of the phosphorylated microtubule-associated protein tau in developing cortical neurons. Neuroscience 1994; 63:895-909.
13. Kanemaru K, Takio K, Miura R et al. Fetal-type phosphorylation of the tau in paired helical filaments. J Neurochem 1992; 58:1667-1675.
14. Goedert M, Jakes R, Crowther RA et al. The abnormal phosphorylation of tau protein at Ser-[202] in Alzheimer disease recapitulates phosphorylation during development. Proc Natl Acad Sci USA 1993; 90:5066-5070.
15. Pope WB, Lambert MP, Leypold B et al. Microtubule-associated protein tau is hyperphopshorylated during mitosis in the human neuroblastoma cell line SH-SY5Y. Exp Neurol 1994; 126:185-194.
16. Ledesma MD, Correas I, Avila J et al. Implication of brain cdc2 and MAP2 kinases in the phosphorylation of tau protein in Alzheimer's disease. FEBS Lett 1992; 308:218-224.
17. Baumann K, Mandelkow EM, Biernat J et al. Abnormal Alzheimer-like phosphorylation of tau-protein by cyclin-dependent kinases cdk2 and cdk5. FEBS Lett 1993; 336:417-424.
18. Arendt T, Holzer M, Grossmann A et al. Increased expression and subcellular translocation of the mitogen activated protein kinase kinase and mitogen-activated protein kinase in Alzheimer's disease. Neuroscience 1995; 68:5-18.
19. Arendt T, Rodel L, Gartner U et al. Expression of the cyclin-dependent kinase inhibitor p16 in Alzheimer's disease. Neuroreport 1996; 7:3047-3049.
20. Vincent I, Rosado M, Davies P. Mitotic mechanisms in Alzheimer's disease? J Cell Biol 1996; 132:413-425.
21. Nagy Z, Esiri MM, Smith AD. Expression of cell division markers in the hippocampus in Alzheimer's disease and other neurodegenerative conditions. Acta Neuropathol (Berl) 1997; 93:294-300.
22. Nagy Z, Esiri MM, Cato AM et al. Cell cycle markers in the hippocampus in Alzheimer's disease. Acta Neuropathol (Berl) 1997; 94:6-15.
23. Zhu X, Rottkamp CA, Raina AK et al. Neuronal CDK7 in hippocampus is related to aging and Alzheimer disease. Neurobiol Aging 2000; 21:807-813.
24. Selkoe DJ. Alzheimer's disease results from the cerebral accumulation and cytotoxicity of amyloid β-protein: A reanalysis of a therapeutic hypothesis. J Alzheimer's Disease 2001; 3:75-81.
25. Ledoux S, Rebai N, Dagenais A et al. Amyloid precursor protein in peripheral mononuclear cells is up-regulated with cell activation. J Immunol 1993; 150:5566-5575.
26. Whitson JS, Selkoe DJ, Cotman CW. Amyloid beta protein enhances the survival of hippocampal neurons in vitro. Science 1989; 243:1488-1490.

27. McDonald DR, Bamberger ME, Combs CK et al. β-amyloid fibrils activate parallel mitogen-activated protein kinase pathways in microglia and THP1 monocytes. J Neurosci 1998; 18:4451-4460.
28. Pyo H, Jou I, Jung S et al. Mitogen-activated protein kinases activated by lipopolysaccharide and beta-amyloid in cultured rat microglia. Neuroreport 1998; 9:871-874.
29. Rapoport M, Ferreira A. PD98059 prevents neurite degeneration induced by fibrillar β-amyloid in mature hippocampal neurons. J Neurochem 2000; 74:125-133.
30. Obrenovich ME, Joseph JA, Atwood CS et al. Amyloid-β: a (life) preserver for the brain. Neurobiol Aging, in press.
31. Giovanni A, Wirtz-Brugger F, Keramaris E et al. Involvement of cell cycle elements, cyclin-dependent kinases, pRb, and E2F x DP, in β-amyloid-induced neuronal death. J Biol Chem 1999; 274:19011-19016.
32. McShea A, Harris PLR, Webster KR et al. Abnormal expression of the cell cycle regulators P16 and CDK4 in Alzheimer's disease. Am J Pathol 1997; 150:1933-1939.
33. McShea A, Zelasko DA, Gerst JL et al. Signal transduction abnormalities in Alzheimer's disease: evidence of a pathogenic stimuli. Brain Res 1999; 815:237-242.
34. McShea A, Wahl AF, Smith MA. Re-entry into the cell cycle: a mechanism for neurodegeneration in Alzheimer disease. Med Hypotheses 1999; 52:525-527.
35. Busser J, Geldmacher DS, Herrup K. Ectopic cell cycle proteins predict the sites of neuronal cell death in Alzheimer's disease brain. J Neurosci 1998; 18:2801-2807.
36 Zhu X, Raina AK, Boux H et al Activation of oncogenic pathways in degenerating neurons in Alzheimer disease. Int J Devl Neurosci 2000; 18.433 437.
37. Zhu X, Rottkamp CA, Boux H et al. Activation of p38 pathway links tau phosphorylation, oxidative stress and cell cycle related events in Alzheimer disease. J Neuropathol Exp Neurol 2000; 59:880-888.
38. Zhu X, Castellani RJ, Takeda A et al. Differential activation of neuronal ERK, JNK/SAPK and p38 in Alzheimer disease: the "two hit" hypothesis. Mech Ageing Dev 2001; 123:39-46.
39. Nagafuchi A, Takeichi M. Transmembrane control of cadherin-mediated cell adhesion: a 94 kDa protein functionally associated with a specific region of the cytoplasmic domain of E-cadherin. Cell Regulation 1989; 1:37-44.
40. Li J, Xu M, Zhou H et al. Alzheimer presenilins in the nuclear membrane, interphase kinetochores, and centrosomes suggest a role in chromosome segregation. Cell 1997; 90:917-927.
41. Wong PC, Zheng H, Chen H et al. Presenilin 1 is required for Notch1 and DII1 expression in the paraxial mesoderm. Nature 1997; 387:288-292.
42. Struhl G, Greenwald I. 1999. Presenilin is required for activity and nuclear access of Notch in Drosophila. Nature 398:522-525.
43. Ye Y, Lukinova N, Fortini ME. Neurogenic phenotypes and altered Notch processing in Drosophila Presenilin mutants. Nature 1999; 398:525-529.
44. Wodarz A, Nusse R. Mechanisms of Wnt signaling in development. Annu Rev Cell Dev Biol 1998; 14:59-88.
45. Polakis P. The oncogenic activation of beta-catenin. Curr Opion Genetics Dev 1999; 9:15-21.
46. Janicki SM, Monteiro MJ. Presenilin overexpression arrests cells in the G1 phase of the cell cycle. Arrest potentiated by the Alzheimer's disease PS2(N141I)mutant. Am J Pathol 1999; 155:135-144.
47. Mattson MP, Guo Q, Furukawa K et al. Presenilins, the endoplasmic reticulum, and neuronal apoptosis in Alzheimer's disease. J Neurochem 1998; 70:1-14.
48. Wolozin B, Iwasaki K, Vito P et al Participation of presenilin 2 in apoptosis; enhanced basal activity conferred by an Alzheimer mutation. Science 1996; 274:1710-1713.
49. Zhang Z, Hartmann H, Do VM et al. Destabilization of beta-catenin by mutations in presenilin-1 potentiates neuronal apoptosis. Nature 1998; 395:698-702.
50. Katayama Y, Kamiya T, Katsura K et al. Studies on brain pyruvate dehydrogenase (PDH) activity and energy metabolites during ischemia and reperfusion. Rinsho Shinkeigaku 1999; 39:1300-1302.
51. Kranenburg O, van der Eb AJ, Zantema A. Cyclin D1 is an essential mediator of apoptotic neuronal cell death. EMBO J 1996; 15:46-54.
52. Park DS, Morris EJ, Padmanabhan J et al. Cyclin-dependent kinases participate in death of neurons evoked by DNA-damaging agents. J Cell Biol 1998; 143:457-467.
53. Evers BM, Ko TC, Li J et al. Cell cycle protein suppression and p21 induction in differentiating Caco-2 cells. Am J Physiol 1996; 271:G722-G727.

54. Greenberg SM, Koo EH, Selkoe DJ et al. Secreted beta-amyloid precursor protein stimulates mitogen-activated protein kinase and enhances tau phosphorylation. Proc Natl Acad Sci USA 1994; 91:7104-7108.

55. Gartner U, Holzer M, Arendt T. Elevated expression of p21ras is an early event in Alzheimer's disease and precedes neurofibrillary degeneration. Neuroscience 1999; 91:1-5.

56. Smith TW, Lippa CF. Ki-67 immunoreactivity in Alzheimer's disease and other neurodegenerative disorders. J Neuropathol Exp Neurol 1995; 54:297-303.

57. Nagy ZS, Smith MZ, Esiri MM et al. Hyperhomocysteinaemia in Alzheimer's disease and expression of cell cycle markers in the brain. Neurol Neurosurg Psychiatry 2000; 69:565-566.

58. Elledge SJ. Cell cycle checkpoints: preventing an identity crisis. Science 1996:274:1664-1672.

59. Stefanis L, Park DS, Yan CY et al. Induction of CPP32-like activity in PC12 cells by withdrawal of trophic support. Dissociation from apoptosis. J Biol Chem 1996; 271:30663-30671.

60. Park DS, Levine B, Ferrari G et al. Cyclin dependent kinase inhibitors and dominant negative cyclin dependent kinase 4 and 6 promote survival of NGF-deprived sympathetic neurons. J Neurosci 1997; 17:8975-8983.

61. Park DS, Obeidat A, Giovanni A et al. Cell cycle regulators in neuronal death evoked by excitotoxic stress: implications for neurodegeneration and its treatment. Neurobiol Aging 2000; 21:771-781.

62. Padmanabhan J, Park DS, Greene LA et al. Role of cell cycle regulatory proteins in cerebellar granule neuron apoptosis. J Neurosci 1999; 19:8747-8756.

63. Koliatsos VE. Biological therapies for Alzheimer's disease: focus on trophic factors. Crit Rev Neurobiol 1996; 10:205-238.

64. Gonzalez AM, Buscaglia M, Ong M et al. Distribution of basic fibroblast growth factor in the 18-day rat fetus: localization in the basement membranes of diverse tissues. J Cell Biol 1990; 110:753-765.

65. Stopa EG, Gonzalez AM, Chorsky R et al. Basic fibroblast growth factor in Alzheimer's disease. Biochem Biophys Res Commun 1990; 171:690-696.

66. Crutcher KA, Scott SA, Liang S et al. Detection of NGF-like activity in human brain tissue: increased levels in Alzheimer's disease. J Neurosci 1993; 13:2540-2550.

67. van der Wal EA, Gomez-Pinilla F, Cotman CW. Transforming growth factor-beta 1 is in plaques in Alzheimer and Down pathologies. Neuroreport 1993; 4:69-72.

68. Lesort M, Johnson GV. Insulin-like growth factor-1 and insulin mediate transient site-selective increases in tau phosphorylation in primary cortical neurons. Neuroscience 2000; 99:305-316.

69. Russell JW, Windebank AJ, Schenone A et al. Insulin-like growth factor-I prevents apoptosis in neurons after nerve growth factor withdrawal. J Neurobiol 1998; 36:455-467.

70. Luth HJ, Holzer M, Gertz HJ et al. Aberrant expression of nNOS in pyramidal neurons in Alzheimer's disease is highly co-localized with p21ras and p16INK4a. Brain 2000; 852:45-55.

71. Vincent I, Jicha G, Rosado M et al. Aberrant expression of mitotic cdc2/cyclin B1 kinase in degenerating neurons of Alzheimer's disease brain. J Neurosci 1997; 17:3588-3598.

72. Arendt T, Holzer M, Gartner U. Neuronal expression of cycline dependent kinase inhibitors of the INK4 family in Alzheimer's disease. J Neural Transm 1998; 105:949-960.

73. Arendt T, Holzer M, Gartner U et al. Aberrancies in signal transduction and cell cycle related events in Alzheimer's disease. J Neural Transm Suppl 1998; 54:147-158.

74. Smith MZ, Nagy Z, Esiri MM. Cell cycle-related protein expression in vascular dementia and Alzheimer's disease. Neurosci Lett 1999; 271:45-48.

75. Tsujioka Y, Takahashi M, Tsuboi Y et al. Localization and expression of cdc2 and cdk4 in Alzheimer brain tissue. Dement Geriatr Cogn Disord 1999; 10:192-198.

76. Tanaka T, Zhong J, Iqbal K et al. The regulation of phosphorylation of tau in SY5Y neuroblastoma cells: the role of protein phosphatases. FEBS Lett 1998; 426:248-254.

77. Sengupta A, Wu Q, Grundke-Iqbal I et al. Potentiation of GSK-3-catalyzed Alzheimer-like phosphorylation of human tau by cdk5. Mol Cell Biochem 1997; 167:99-105.

78. Pei JJ, Grundke-Iqbal I, Iqbal K et al. Accumulation of cyclin-dependent kinase 5 (cdk5) in neurons with early stages of Alzheimer's disease neurofibrillary degeneration. Brain Res 1998; 797:267-277.

79. Patrick GN, Zukerberg L, Nikolic M et al. Conversion of p35 to p25 deregulates Cdk5 activity and promotes neurodegeneration. Nature 1999; 402:615-622.

80. Flaherty DB, Soria JP, Tomasiewicz HG et al. Phosphorylation of human tau protein by microtubule-associated kinases: GSK3beta and cdk5 are key participants. J Neurosci Res 2000; 62:463-472.

81. Bennecib M, Gong CX, Grundke-Iqbal I et al. Role of protein phosphatase-2A and -1 in the regulation of GSK-3, cdk5 and cdc2 and the phosphorylation of tau in rat forebrain. FEBS Letters 2000; 485:87-93.

82. Takahashi M, Iseki E, Kosaka K. Cdk5 and munc-18/p67 co-localization in early stage neurofibrillary tangles-bearing neurons in Alzheimer type dementia brains. J Neurol Sci 2000; 172:63-69.

83. Ahlijanian MK, Barrezueta NX, Williams RD et al. Hyperphosphorylated tau and neurofilament and cytoskeletal disruptions in mice overexpressing human p25, an activator of cdk5. Proc Natl Acad Sci USA 2000; 97:2910-2915.

84. Lee KY, Clark AW, Rosales JL et al. Elevated neuronal Cdc2-like kinase activity in the Alzheimer disease brain. Neurosci Res 1999; 34:21-29.

85. Gartner U, Holzer M, Heumann R et al. Induction of p21ras in Alzheimer pathology. Neuroreport 1995; 6:1441-1444.

86. Raina AK, Pardo P, Rottkamp CA et al. Senescent emergence of neurons in Alzheimer disease. Mech Ageing Dev 2001; 123:3-9.

87. Masliah E, Mallory M, Alford M et al. Immunoreactivity of the nuclear antigen p105 is associated with plaques and tangles in Alzheimer's disease. Lab Invest 1993; 69:562-569.

88. Kitamura Y, Shimohama S, Kamoshima W et al. Changes of p53 in the brains of patients with Alzheimer's disease. Biochem Biophys Res Commun 1997; 232:418-421.

89. de la Monte SM, John YK, Wands JR. Correlates of p53- and Fas (CD95)-mediated apoptosis in Alzheimer's disease. J Neurol Sci 1997; 152:73-83.

90. Shimohama S, Kamiya S, Taniguchi T et al. Differential involvement of small G proteins in Alzheimer's disease. Int J Mol Med 1999; 3:597-600.

91. Culvenor JG, Evin G, Cooney MA et al. Presenilin 2 expression in neuronal cells: induction during differentiation of embryonic carcinoma cells. Exp Cell Res 2000; 255:192-206.

92. North S, El-Ghissassi F, Pluquet O et al. The cytoprotective aminothiol WR1065 activates p21waf-1 and down regulates cell cycle progression through a p53-dependent pathway. Oncogene 2000; 19:1206-1214.

93. Smits VA, Klompmaker R, Arnaud L et al. Polo-like kinase-1 is a target of the DNA damage checkpoint. Cell Biol 2000; 2:672-676.

94. Harris PLR, Zhu X, Pamies C et al. Neuronal polo-like kinase in Alzheimer disease indicates cell cycle changes. Neurobiol Aging 2000; 21:837-841.

95. Drewes G, Mandelkow EM, Baumann K et al. Dephosphorylation of tau protein and Alzheimer paired helical filaments by calcineurin and phosphatase-2A. FEBS Lett 1993; 336:425-432.

96. Trojanowski JQ, Lee VW. Phosphorylation of paired helical filament tau in Alzheimer's disease neurofibrillary lesions; focusing on phosphatases. FASEB J 1995; 9:1570-1576.

97. Garver TD, Lehman RA, Billingsley ML. Microtubule assembly competence analysis of freshly-biopsied human tau, dephosphorylated tau, and Alzheimer tau. J Neurosci Res 1996; 44:12-20.

98. Sontag E, Nunbhakdi-Craig V, Lee G et al. Molecular interactions among protein phosphatase 2A, tau, and microtubules. Implications for the regulation of tau phosphorylation and the development of tauopathies. J Biol Chem 1999; 274:25490-25498.

99. Husseman JW, Nochlin D, Vincent I. Mitotic activation: a convergent mechanism for a cohort of neurodegenerative diseases. Neurobiol Aging 2000; 21:815-828.

100. Ding XL, Husseman J, Tomashevski A et al. The cell cycle Cdc25A tyrosine phosphatase is activated in degenerating postmitotic neurons in Alzheimer's disease. Am J Pathol 2000; 157:1983-1990.

101. Trejo J, Massamiri T, Deng T et al. A direct role for protein kinase C and the transcription factor Jun/AP-1 in the regulation of the Alzheimer's beta-amyloid precursor protein gene. J Biol Chem 1994; 269:21682-21690.

102. Singh TJ, Grundke-Iqbal I, McDonald B et al. Comparison of the phosphorylation of microtubule-associated protein tau by non-proline dependent protein kinases. Mol Cell Biochem 1994; 131:181-189.

103. Suzuki T, Ando K, Isohara T et al. Phosphorylation of Alzheimer beta-amyloid precursor-like proteins. Biochemistry 1997; 36:4643-4649.

104. Lanius RA, Wagey R, Sahl B et al. Protein kinase C activity and protein levels in Alzheimer's disease. Brain Res 1997; 764:75-80.

105. Nakai M, Hojo K, Taniguchi, T et al. PKC and tyrosine kinase involvement in amyloid beta (25-35)-induced chemotaxis of microglia. Neuroreport 1998; 9:3467-3470.

106. Favit A, Grimaldi M, Nelson TJ et al. Alzheimer's-specific effects of soluble beta-amyloid on protein kinase C-alpha and -gamma degradation in human fibroblasts. Proc Natl Acad Sci USA 1998; 95:5562-5567.

107. Bhagavan S, Ibarreta D, Ma D et al. Restoration of TEA-induced calcium responses in fibroblasts from Alzheimer's disease patients by a PKC activator. Neurobiol Disease 1998; 5:177-187.

108. Gargiulo L, Bermejo M, Liras A. Reduced neuronal nitric oxide synthetase and c-protein kinase levels in Alzheimer's disease. Rev Neurologia 2000; 30:301-303.

109. Tsujio I, Tanaka T, Kudo T et al. Inactivation of glycogen synthase kinase-3 by protein kinase C delta: implications for regulation of tau phosphorylation. FEBS Lett 2000; 469:111-117.

110. Singh TJ, Zaidi T, Grundke-Iqbal I et al. Non-proline-dependent protein kinases phosphorylate several sites found in tau from Alzheimer disease brain. Mol Cell Biochem 1996; 154:143-151.

111. Marambaud P, Ancolio K, Alves da Costa C et al. Effect of protein kinase A inhibitors on the production of Abeta40 and Abeta42 by human cells expressing normal and Alzheimer's disease-linked mutated betaAPP and presenilin 1. Brit J Pharmacol 1999; 126:1186-1190.

112. Kawamata T, Taniguchi T, Mukai H et al. A protein kinase, PKN, accumulates in Alzheimer neurofibrillary tangles and associated endoplasmic reticulum-derived vesicles and phosphorylates tau protein. J Neurosci 1998; 18:7402-7410.

113. Hong M, Chen DC, Klein PS et al. Lithium reduces tau phosphorylation by inhibition of glycogen synthase kinase-3. J Biol Chem 1997; 272:25326-25332.

114. Zubenko GS, Stiffler JS, Hughes HB et al. Reductions in brain phosphatidylinositol kinase activities in Alzheimer's disease. Biol Psychiatry 1999; 45:731-736.

115. Tanaka T, Tsujio I, Nishikawa T et al. Significance of tau phosphorylation and protein kinase regulation in the pathogenesis of Alzheimer disease. Alzheimer Dis Assoc Disord 2000; 14(Suppl 1):S18-S24.

116. Yuan J, Yankner BA. Apoptosis in the nervous system. Nature 2000; 407:802-809.

117. Stadelmann C, Bruck W, Bancher C et al. Alzheimer disease: DNA fragmentation indicates increased neuronal vulnerability, but not apoptosis. J Neuropathol Exp Neurol 1998; 57:456-464.

118. Weihl CC, Ghadge GD, Kennedy SG et al. Mutant presenilin-1 induces apoptosis and downregulates Akt/PKB. J Neurosci 1999; 19:5360-5369.

119. Anderton BH, Dayanandan R, Killick R et al. Does dysregulation of the Notch and wingless/Wnt pathways underlie the pathogenesis of Alzheimer's disease? Mol Med Today 2000; 6:54-59.

120. Luedecking EK, DeKosky ST, Mehdi H et al. Analysis of genetic polymorphisms in the transforming growth factor-beta1 gene and the risk of Alzheimer's disease. Human Genetics 2000; 106:565-569.

121. Luterman JD, Haroutunian V, Yemul S et al. Cytokine gene expression as a function of the clinical progression of Alzheimer disease dementia. Arch Neurol 2000; 57:1153-1160.

122. Lu Q, Soria JP, Wood JG. p44mpk MAP kinase induces Alzheimer type alterations in tau function and in primary hippocampal neurons. J Neurosci Res 1993; 35:439-444.

123. Mandelkow EM, Biernat J, Drewes G et al. Microtubule-associated protein tau, paired helical filaments, and phosphorylation. Ann NY Acad Sci 1993; 695:209-216.

124. Hyman BT, Elvhage TE, Reiter J. Extracellular signal regulated kinases. Localization of protein and mRNA in the human hippocampal formation in Alzheimer's disease. Am J Pathol 1994; 144:565-572.

125. Torack RM, Miller JW. Denervation induced abnormal phosphorylation in hippocampal neurons. Brain Res 1995; 669:135-139.

126. Mills J, Laurent Charest D, Lam F et al. Regulation of amyloid precursor protein catabolism involves the mitogen-activated protein kinase signal transduction pathway. J Neurosci 1997; 17:9415-9422.

127. Luo Y, Bond JD, Ingram VM. Compromised mitochondrial function leads to increased cytosolic calcium and to activation of MAP kinases. Proc Natl Acad Sci USA 1997; 94:9705-9710.

128. Perry G, Roder H, Nunomura A et al. Activation of neuronal extracellular receptor kinase (ERK) in Alzheimer disease links oxidative stress to abnormal phosphorylation. Neuroreport 1999; 10:2411-2415.
129. Ekinci FJ, Shea TB. Hyperactivation of mitogen-activated protein kinase increases phospho-tau immunoreactivity within human neuroblastoma: additive and synergistic influence of alteration of additional kinase activities. Cell Mol Neurobiol 1999; 19:249-260.
130. Knowles RB, Chin J, Ruff CT et al. Demonstration by fluorescence resonance energy transfer of a close association between activated MAP kinase and neurofibrillary tangles: implications for MAP kinase activation in Alzheimer disease. J Neuropathol Exp Neurol 1999; 58:1090-1098.
131. Toran-Allerand CD, Singh M, Setalo G. Novel mechanisms of estrogen action in the brain: new players in an old story. Frontiers Neuroendocrinol 1999; 20:97-121.
132. Takeda A, Perry G, Abraham NG et al. Overexpression of heme oxygenase in neuronal cells, the possible interaction with tau. J Biol Chem 2000; 275:5395-5399.
133. Reynolds CH, Nebreda AR, Gibb GM et al. Reactivating kinase/p38 phosphorylates tau protein in vitro. J Neurochem 1997; 69:191-198.
134. Reynolds CH, Betts JC, Blackstock WP. Phosphorylation sites on tau identified by nanoelectrospray mass spectrometry: differences in vitro between the mitogen-activated protein kinases ERK2, c-Jun N-terminal kinase and P38, and glycogen synthase kinase-3beta. Neurochem 2000; 74:1587-1595.
135. Guise S, Braguer D, Carles G et al. Hyperphosphorylation of tau is mediated by ERK activation during anticancer drug-induced apoptosis in neuroblastoma cells. J Neurosci Res 2001; 63:257-267.
136. Steiner B, Mandelkow EM, Biernat J et al. Phosphorylation of microtubule-associated protein tau: identification of the site for Ca2(+)-calmodulin dependent kinase and relationship with tau phosphorylation in Alzheimer tangles. EMBO J 1990; 9:3539-3544.
137. Utton MA, Vandecandelaere A, Wagner U et al. Phosphorylation of tau by glycogen synthase kinase 3beta affects the ability of tau to promote microtubule self-assembly. Biochem J 1997; 323:741-747.
138. Floyd RA. Antioxidants, oxidative stress, and degenerative neurological disorders. Proc Soc Exp Biol Med 1999; 222:236-245.
139. Hensley K, Floyd RA, Zheng NY et al. p38 kinase is activated in the Alzheimer's disease brain. J Neurochem 1999; 72:2053-2058.
140. Zhu X, Raina AK, Rottkamp CA et al. Activation and redistribution of c-Jun N-terminal kinase/ stress activated protein kinase in degenerating neurons in Alzheimer's disease. J Neurochem 2001; 76:435-441.
141. Mandelkow EM, Drewes G, Biernat J et al. Glycogen synthase kinase-3 and the Alzheimer-like state of microtubule-associated protein tau. FEBS Lett 1992; 314:315-321.
142. Hanger DP, Hughes K, Woodgett JR et al. Glycogen synthase kinase-3 induces Alzheimer's disease-like phosphorylation of tau: generation of paired helical filament epitopes and neuronal localisation of the kinase. Neurosci Lett 1992; 147:58-62.
143. Yang SD, Yu JS, Liu WK et al. Synergistic control mechanism for abnormal site phosphorylation of Alzheimer's diseased brain tau by kinase FA/GSK-3 alpha. Biochem Biophys Res Commun 1993; 197:400-406.
144. Yang SD, Yu JS, Shiah SG et al. Protein kinase FA/glycogen synthase kinase-3 alpha after heparin potentiation phosphorylates tau on sites abnormally phosphorylated in Alzheimer's disease brain. J Neurochem 1994; 63:1416-1425.
145. Singh TJ, Haque N, Grundke-Iqbal I et al. Rapid Alzheimer like phosphorylation of tau by the synergistic actions of non-proline-dependent protein kinases and GSK-3. FEBS Lett 1995; 358:267-272.
146. Hoshi M, Takashima A, Noguchi K et al. 1996. Regulation of mitochondrial pyruvate dehydrogenase activity by tau protein kinase I/glycogen synthase kinase 3beta in brain. Proc Natl Acad Sci USA 1996; 93:2719-2723.
147. Yamaguchi H, Ishiguro K, Uchida T et al. Preferential labeling of Alzheimer neurofibrillary tangles with antisera for tau protein kinase (TPK) I/glycogen synthase kinase-3 beta and cyclin-dependent kinase 5, a component of TPK II. Acta Neuropathol (Berl) 1996; 92:232-241.
148. Pei JJ, Tanaka T, Tung YC et al. Distribution, levels, and activity of glycogen synthase kinase-3 in the Alzheimer disease brain. J Neuropathol Exp Neurol 1997; 56:70-78.

149. Wang JZ, Wu Q, Smith A et al. Tau is phosphorylated by GSK-3 at several sites found in Alzheimer disease and its biological activity markedly inhibited only after it is prephosphorylated by A-kinase. FEBS Lett 1998; 436:28-34.

150. Gantier R, Gilbert D, Dumanchin C et al. The pathogenic L392V mutation of presenilin 1 decreases the affinity to glycogen synthase kinase-3 beta. Neurosci Lett 2000; 283:217-220.

151. Van Gassen G, De Jonghe C, Nishimura M et al. Evidence that the beta-catenin nuclear translocation assay allows for measuring presenilin 1 dysfunction. Mol Med 2000; 6:570-580.

152. Baki L, Marambaud P, Efthimiopoulos S et al. Presenilin-1 binds cytoplasmic epithelial cadherin, inhibits cadherin/p120 association, and regulates stability and function of the cadherin/catenin adhesion complex. Proc Natl Acad Sci USA 2001; 98:2381-2386.

153. Anderson AJ, Cummings BJ, Cotman CW. Increased immunoreactivity for Jun- and Fos-related proteins in Alzheimer's disease: association with pathology. Exp Neurol 1994; 125:286-295.

154. Layfield R, Fergusson J, Aitken A et al. Neurofibrillary tangles of Alzheimer's disease brains contain 14-3-3 proteins. Neurosci Lett 1996; 209:57-60.

155. Hashiguchi M, Sobue K, Paudel HK. 14-3-3zeta is an effector of tau protein phosphorylation. J Biol Chem 2000; 275:25247-25254.

156. Anderson AJ, Su JH, Cotman CW. DNA damage and apoptosis in Alzheimer's disease: colocalization with c-Jun immunoreactivity, relationship to brain area, and effect of postmortem delay. J Neurosci 1996; 16:1710-1719.

157. Ferrer I, Segui J, Planas AM. Amyloid deposition is associated with c-Jun expression in Alzheimer's disease and amyloid angiopathy. Neuropathol Appl Neurobiol 1996; 22:521-526.

158. Marcus DL, Strafaci JA, Miller DC et al. Quantitative neuronal c-fos and c-jun expression in Alzheimer's disease. Neurobiol Aging 1998; 19:393-400.

159. Kihiko ME, Tucker HM, Rydel RE et al. c-Jun contributes to amyloid beta-induced neuronal apoptosis but is not necessary for amyloid beta-induced c-jun induction. J Neurochem 1999; 73:2609-2612.

160. Shirazi SK, Wood JG. The protein tyrosine kinase, fyn, in Alzheimer's disease pathology. Neuroreport 1993; 4:435-437.

161. Lambert MP, Barlow AK, Chromy BA et al. Diffusible, nonfibrillar ligands derived from Abeta1-42 are potent central nervous system neurotoxins. Proc Natl Acad Sci USA 1998; 95:6448-6453.

162. Lee C, Kim MG, Jeon SH et al. Two species of mRNAs for the fyn proto-oncogene are produced by an alternative polyadenylation. Molecules and Cell 1998; 8:746-749.

163. Chopp M. The roles of heat shock proteins and immediate early genes in central nervous system normal function and pathology. Curr Opin Neurol Neurosurg 1993; 6:6-10.

164. Sayas CL, Moreno-Flores MT, Avila J et al. The neurite retraction induced by lysophosphatidic acid increases Alzheimer's disease-like Tau phosphorylation. J Biol Chem 1999; 274:37046-37052.

165. Raina AK, Hochman A, Zhu X et al. Abortive apoptosis in Alzheimer's disease. Acta Neuropathol 2001; 101:305-310.

166. Barni S, Sciola L, Spano A et al. Static cytofluorometry and fluorescence morphology of mitochondria and DNA in proliferating fibroblasts. Biotechnic Histochem 1996; 71:66-70.

167. Hirai K, Aliev G, Nunomura A et al. Mitochondrial abnormalities in Alzheimer's disease. J Neurosci 2001; 21:3017-3023.

168. Sousa M, Barros A, Silva J et al. Developmental changes in calcium content of ultrastructurally distinct subcellular compartments of preimplantation human embryos. Mol Human Reproduction 1997; 3:83-90.

169. Zhu X, Rottkamp CA, Hartzler A et al. Activation of MKK6, an upstream activator of p38, in Alzheimer's disease. J Neurochem 2001; 79:311-318.

170. Zhang L, Jope RS. Oxidative stress differentially modulates phosphorylation of ERK, p38 and CREB induced by NGF or EGF in PC12 cells. Neurobiol Aging 1999; 20:271-278.

171. Smith MA, Sayre LM, Monnier VM et al. Radical AGEing in Alzheimer's disease. Trends Neurosci 1995; 18:172-176.

172. Markesbery WR. Oxidative stress hypothesis in Alzheimer's disease. Free Radic Biol Med 1997; 23:134-147.

173. Curcio F, Ceriello A. Decreased cultured endothelial cell proliferation in high glucose medium is reversed by antioxidants: new insights on the pathophysiological mechanisms of diabetic vascular complications. In Vitro Cell Dev Biol 1992; 28A:787-790.

174. Ferrari G, Yan CY, Greene LA. N-acetylcysteine (D- and L-stereoisomers) prevents apoptotic death of neuronal cells. J Neurosci 1995; 15:2857-2866.

175. Terman A. Garbage catastrophe theory of aging: imperfect removal of oxidative damage? Redox Report 2001; 6:15-26.

176. Perry G, Friedman R, Shaw G et al. Ubiquitin is detected in neurofibrillary tangles and senile plaque neurites of Alzheimer disease brains. Proc Natl Acad Sci USA 1987; 84:3033-3036.

177. Bence NF, Sampat RM, Kopito RR. Impairment of the ubiquitin-proteasome system by protein aggregation. Science 2001; 292:1552-1555.

178. Masaki R, Saito T, Yamada K et al. Accumulation of phosphorylated neurofilaments and increase in apoptosis-specific protein and phosphorylated c-Jun induced by proteasome inhibitors. J Neurosci Res 2000; 62:75-83.

179. King RW, Deshaies RJ, Peters J-M et al. How proteolysis drives the cell cycle. Science 1996; 274:1652-1659.

180. Smith MA, Taneda S, Richey PL et al. Advanced Maillard reaction end products are associated with Alzheimer disease pathology. Proc Natl Acad Sci USA 1994; 91:5710-5714.

181. Reddy VP, Obrenovich MF, Atwood CS et al. Involvement of Maillard reactions in Alzheimer disease. Neurotoxicity Research 2002, 4:191-209.

182. Lee CK, Klopp RG, Weindruch R et al. Gene expression profile of aging and its retardation by caloric restriction. Science 1999; 285:1390-1393.

From Cell Cycle Activation to the Inhibition of the Wnt Pathway:

A Hypothesis of How Neurons Die in Response to β-Amyloid

Agata Copani, Filippo Caraci, Maria Angela Sortino, Ferdinando Nicoletti and Andrea Caricasole

A decade of studies has substantiated the concept that mitotic reactivation is the way neurons die in numerous neurodegenerative disorders.[1-8] Most of us tend to agree with the original hypothesis by Heintz that apoptotic death is the result of a failed attempt of neurons to enter the cell cycle;[9] however, we are unaware of when and how a dividing adult neuron dies. As yet, we do not know precisely where to look for these answers given our poor knowledge of what in the cell cycle of a normally dividing cell applies to a neuron. In this review, we will try to answer some of these mechanistic questions focusing on the ectopic reentrance of differentiated neurons into the cell cycle initiated by β-amyloid (Aβ).[10,11] In doing so, we will address the obscure issue of replicative DNA synthesis in adult neurons, its relationship with DNA damage, tau protein hyperphosphorylation and apoptotic death, and we will finally discuss some pharmacological perspectives.

The Regulation of the Cell Cycle by Aβ

Cell cycle in proliferating cells depends on the sequential activation of cyclin (Cyc)/cyclin dependent protein kinases (CDKs), which control the transition through the different phases of the cycle.[12] Activation of CDK4-6 by the cyclin partner, Cyc D, is involved in the progression of the mid G1 phase, whereas activation of CDK2 by Cyc E and CycA controls G1/S transition, and S phase progression. Finally, the Cyc B/CDK1 complex is the regulator of the G2/M transition.

Aβ activates the cell cycle in neurons by inducing the sequential expression of cell cycle proteins usually functioning in proliferating cells.[10] Following Aβ treatment of rat cortical neurons, we observed the induction of cyclin D1 and the phosphorylation of retinoblastoma (the substrate of Cyc D/CDK4-6 complexes), followed by the induction of cyclin E and A. The most interesting observation was that Aβ-treated neurons started DNA replication before undergoing apoptosis. DNA replication was assessed by fluorescence-activated cell sorter (FACS) analysis of DNA contents, thus eliminating the potential bias of a bromodeoxyuridine labeling of DNA synthesis associated with DNA repair rather than DNA replication. By means of FACS analysis, which allows the simultaneuos evaluation of S phase and apoptosis, we found

Cell Cycle Mechanisms and Neuronal Cell Death, edited by Agata Copani and Ferdinando Nicoletti. ©2005 Eurekah.com and Kluwer Academic / Plenum Publishers.

that blockade of the G1/S transition using a cyclin D1 antisense or a dominant-negative mutant of CDK-2 prevented both Aβ-induced DNA replication and apoptosis.[10]

A number of conclusions can be drawn from these and other studies.[13-15] First, neurons, although postmitotic and highly differentiated cells, express constitutively some cell cycle regulatory elements (e.g., CDKs). They can also be induced to express cell cycle proteins (e.g., cyclins), and perhaps they miss just few essential components of the cell cycle. Second, the reactivation of the cell cycle is an obligatory step in the apoptotic pathway evoked by Aβ. Third, and most important, the ectopic S phase triggers the death of neurons. How? We thought that the answer could stay in the mechanisms that allow neurons to replicate their DNA.

The DNA Replication Machinery Activated by Aβ in Adult Neurons

Sequential activation of DNA polymerase (pol)-α, and -δ or -ε (for a review, see Ref. 16) mediates DNA synthesis in proliferating cells. Polα with its associate primase activity, is unique among DNA pols because of its ability to initiate de novo DNA synthesis. The enzyme synthesizes short RNA-DNA primers for the leading strand synthesis and for each Okazaki fragment on the lagging strand. Polα is constituted by four distinct proteins with molecular weight of 180, 70, 58 and 49 kD. The p180 subunit is the catalytitic subunit of pol-α, while the p49 and p58 subunits form the primase complex. Polδ and/or polε are thought to function as major replicative DNA pols by replacing polα and extending the nascent DNA primers. Both enzymes are endowed with a proofreading exonuclease activity that is suited for replication with minimal errors.

In addition, polα coordinates DNA replication to cell cycle responses in a way that the onset of mitosis is blocked while DNA is being synthesized or damaged. Polδ and polε can function as DNA repair enzymes for the correction of mismatch errors that have escaped the editing process during replication. Thus, the coordinated management of the activity of multiple DNA polymerases is a fundamental aspect of DNA replication in normally dividing cells.

Unlike the replicative DNA pols, certain pols are specialized to bypass with high fidelity specific types of DNA damage occuring during replication. However, these pols produce errors at a high frequency when they operate on undamaged DNA.[17] These specialised pols include polβ, which is the smallest known DNA pol. Polβ partecipates in base excision repair, a repair pathway universally present in living cells for eliminating and replacing oxidised bases.[18]

At present, we lack the exact knowledge of the repertoire of DNA pols expressed by neurons. Most of the informations come from the use of pharmacological inhibitors with a certain degree of selectivity towards individual DNA pols. In isolated neuronal fractions, polβ activity was found to be predominant, although some polδ/ε and a little polα activity were also present. However, the contribution of dividing neuronal precursors to these activities could not be excluded.[19]

Cultured cortical neurons express polδ and its ancillary protein, PCNA, but not polε.[11] Treatment of neurons with toxic concentrations of AβP induces the expression of the p49 and the p58 subunits of the primase/polα complex. Intriguingly, the p180 catalytic subunit of pol-α remains undetectable in neurons treated with AβP. AβP instead induces the expression of the repair enzyme polβ in a cell cycle dependent manner. This is unusual, because expression of pol-β is expected to be regulated by the extent of DNA damage but not by the activation of the cell cycle.[11]

Moreover, the knockdown of polβ prevented Aβ-induced DNA synthesis and the ensuing apoptosis; similar effects were observed by knocking down the p49 primase subunit.[11] These results suggest that AβP activates a primase-directed, but non-canonical, pathway of DNA replication that is mediated by polβ in neurons (Fig. 1).

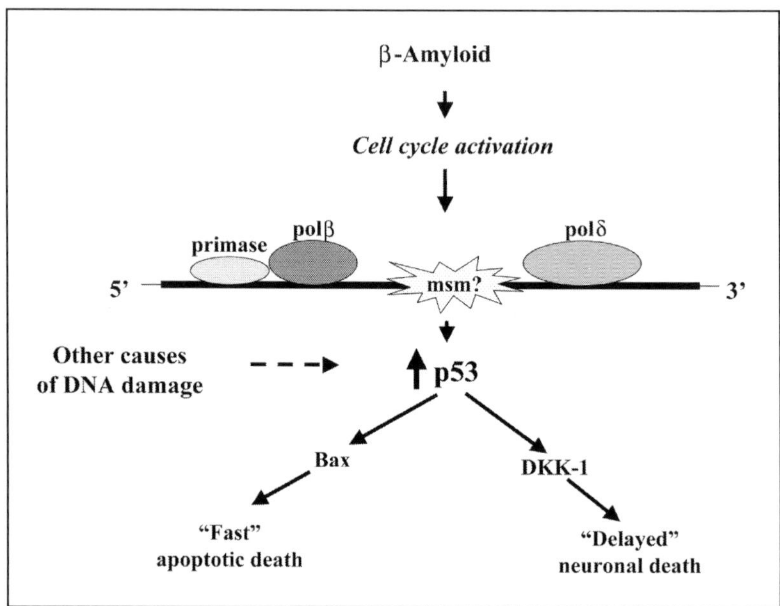

Figure 1. Hypothetical model showing the pathological cascade induced by Aβ in neurons. Extracellular Aβ induces the activation of an unscheduled cell cycle through an unknown mechanism. DNA synthesis results from the sequential involvement of DNA primase, DNA polβ, and DNA polδ. The noncanonical replication carried out by polβ generates mismatches (msm) in the newly generated DNA strand, thus contributing to the overall DNA damage in neurons challenged with Aβ. The ensuing increase in p53 may either promote the execution phase of apoptotic death, via the induction of Bax and other pro-apoptotic factors, or trigger a slow degenerative process *via* the induction of DKK-1 and the resulting inhibition of the Wnt pathway.

Cell Cycle Signaling, DNA Damage and Neuronal Apoptosis

An unanswered question is how the noncanonical pathway of DNA replication initiated by Aβ leads to neuronal apoptosis. Given that polβ is an error-prone enzyme, one possibility is that the aberrant DNA replication induced by Aβ might contribute to the overall DNA damage observed in AD (Alzheimer's disease) neurons[20] (Fig. 1). Accordingly, an increased expression of p53, an established sensor of DNA damage, occurs in neurons challenged with Aβ and is prevented by knocking down polβ.[11] The presence of tetraploid neurons in AD brain[21] suggests that this error-prone replication mechanism somehow proceeds without signaling to cell cycle checkpoints until neurons become loaded with unsustainable DNA damage. In other words, de novo DNA synthesis might be a potential source of replication errors which can contribute to reach the threshold for the activation of a p53/DNA damage-dependent pathway of death in neurons[11] (Fig. 1). Thus, the contribution of the cell cycle to the death of neurons may not be always indispensable, and might depend on the severity of the insult (i.e., strong insults can be sufficient to cause lethal DNA damage) and/or on the vulnerability of neurons.

Waiting to reach the threshold for death, neurons might acquire the features peculiar to a specific disease. To this regard, we found that by slowing down the execution of apoptotic death with a caspase inhibitor we could amplify the phenomenon of tau hyperphosphorylation in AβP-treated neurons. Interestingly, the p53 target gene, dickkopf-1 (DKK-1), was critically involved in the process of tau hyperphorylation[22] (Fig. 1). DKK-1 is a negative modulator of the Wnt signaling pathway, which promotes developmental neuroproliferation and differentiaton[23] and is also required to maintain the survival of mature neurons challenged

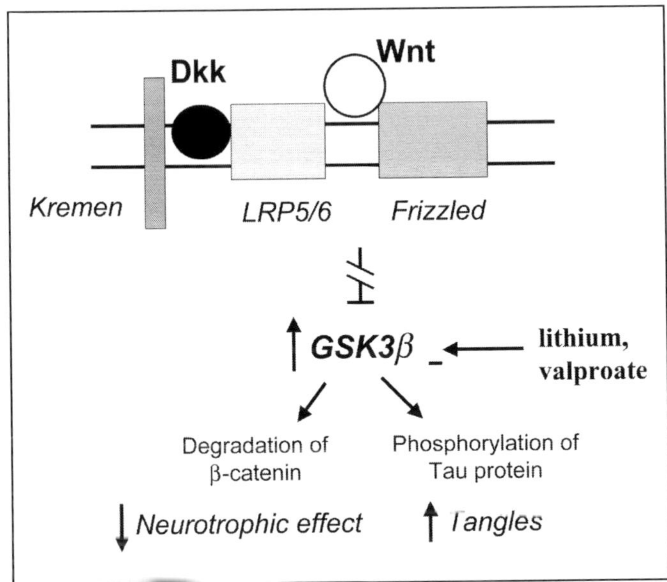

Figure 2. DKK-1 released extracellularly negatively modulates the Wnt pathway by interacting with the Wnt coreceptor LRP5/6, which is then transferred inside the cell and is no longer available to bind Wnt. The increased GSK3β activity (resulting from the inhibition of the Wnt pathway) leads to tau protein hyperphosphorylation and to the degradation of β-catenin. This leads to a reduced expression of neurotrophic genes regulated by the transcriptional activity of β-catenin. Lithium and valproate (divalproex) can overcome the detrimental effects of DKK-1 by inhibiting GSK3β activity.

with Aβ.[24,25] Activation of this pathway leads to the inhibition of the putative tau-phosphorylating enzyme glycogen synthase kinase-3β (GSK-3β) by dissociating the enzyme from a multiprotein complex that involves axin, adenomatous polyposis coli, and β-catenin.[26] By inhibiting GSK3β, the Wnt signaling pathway precludes the formation of neurofibrillary tangles. Inhibition of GSK3β also prevents phosphorylation of β-catenin, which thus escapes degration and translocates to the nucleus where it drives the expression of Wnt-target survival genes, such as Bcl-2[27] (Fig. 2). We found that DKK-1 expression in Aβ-treated neurons was dependent on p53, and that the knock-down of DKK-1 nearly abolished tau hyperphosphorylation. In AD brain, DKK-1 was present in neurons expressing p53 and it was colocalized with neurofibrillary tangles. Thus, it appears that Aβ triggers a pathological cascade in neurons by activating an unscheduled cell cycle leading to an abnormal DNA synthesis carried out by the "noncanonical" enzyme DNA polβ. DNA damage may be generated by this and additional factors, including excitotoxicity and inflammation (Fig. 1). The resulting increase in p53 expression might be viewed as a central element in DNA repair (in surviving neurons), execution of apoptotic death (in neurons that are rapidly killed by Aβ), and tau hyperphosphorylation (in neurons that die slowly as a result of cytoskeletal derangement) (Fig. 1). The fate of the latter population of neurons will be critically depend on the efficiency of additional rescue pathways that can compensate for the DKK-1-induced inhibition of the Wnt pathway. Among these, the phosphatidylinositol-3-kinase/Akt pathway activated by membrane tyrosine kinase receptors (such as insulin, bFGF or neurotrophin receptors) is a likely candidate because it negatively regulates the activity of GSK3β through a phosphorylation mechanism mediated by phosphorylated-Akt (Fig. 3). It is particularly interesting that polymorphic variations of the p85α subunit of the phosphatidylinositol-3-kinase might increase the risk of developing late-onset AD.[28]

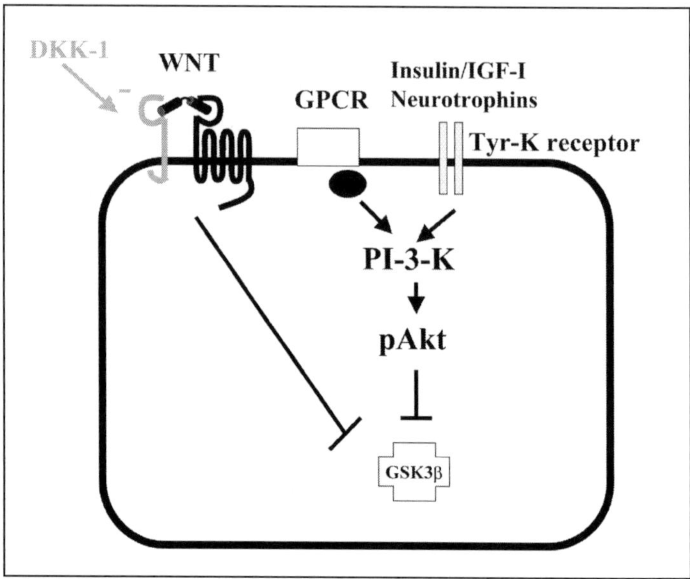

Figure 3. The overall sensitivity of vulnerable neurons to the inhibition of the Wnt pathway is critically influenced by complementary receptor signaling pathways leading to the inhibition of GSK3β. Both G-protein coupled receptors (GCPRs) and tyrosin kinase (Tyr-K) receptors may control the activity of GSK3β through the phosphatidylinositol-3-kinase (PI-3-K)/Akt pathway. Stimulation of the PI-3-K pathway by activated GCPRs is mediated by the βγ subunit of the G protein. Tyr-K receptors activated by insulin, insulin-like growth factor-I (IGF-I), neurotrophins, and other trophic factors, stimulate PI-3-K through tyrosin phosphorylation mechanisms. Activated (phosphorylated) Akt phosphorylates GSK3β at Ser 9, inhibiting the enzyme activity. A defective functioning of insulin receptors or other "rescuing" receptors will increase neuronal susceptibility to the detrimental effect of DKK-1, facilitating the formation of neurofibrillary tangles.

Pharmacological Perspectives

In an effort to put the management of the cell cycle into a therapeutic perspective, we must move from evidence obtained in cultured neurons exposed to Aβ. Exogenous transforming growth factor-β (TGF-β) protects neurons against Aβ by halting the cell cycle.[10] Estrogens enhance the secretion of TGF-β from astrocytes in culture, and the collected medium protect neurons against Aβ toxicity[29] (Fig. 4). Thus, one possible strategy would be to promote cell cycle arrest by inducing the local production of endogenous cytostatic factors such as TGF-β. Agonists of group-II metabotropic glutamate receptors (i.e., mGlu2 and -3) increase the production of TGFβ in cultured astrocytes[30-32] and protect neurons surrounded by astrocytes against Aβ toxicity[33] (Fig. 4).

Cytostatic drugs that act as CDK inhibitors (e.g., mimosine and flavopiridol) are protective against Aβ-induced toxicity;[10] however, the adverse effects of these drugs (that are proper to all antineoplastic agents) might seriously limit their use.

Since the repair enzyme polβ carries out a large component of DNA replication and apoptotic death in neurons exposed to Aβ, specific inhibitors of this enzyme, such as the natural glycoside prunasin, should be tested for their ability to prevent Aβ toxicity. It is likely that selective inhibitors of polβ might represent a key to neuronal-specific cell cycle inhibition as opposed to classical antineoplastic agents.

Finally, drugs that mimic the Wnt pathway by inhibiting GSK3β, such as lithium ions,[34] divalproex, or synthetic enzyme inhibitors, are promising for their potential ability to prevent

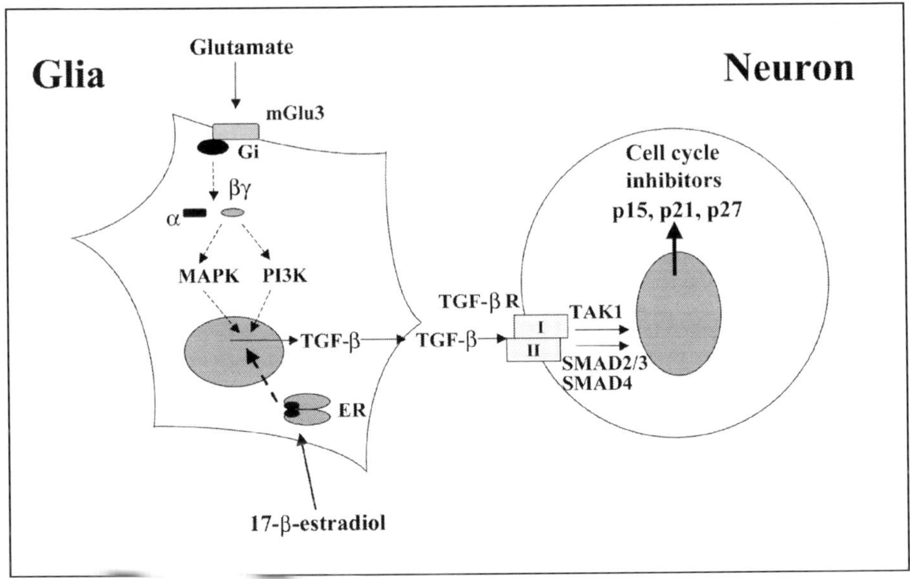

Figure 4. Agents that are known to protect neurons against Aβ toxicity, such as mGlu2/3 receptor agonists and estrogens may stimulate astrocytes to produce transforming growth factor-β (TGFβ) in addition to other neuroprotective factors. TGFβ would act on neighbour neurons by inducing cell cycle arresting factors (i.e., p15, p21, and p27), which inhibit the activity of CDKs, thus halting the progression of the cycle initiated by Aβ (modified from J Cereb Blood Flow Metab 2001; 21:1013-1033).

tau hyperphosphorylation and neurofibrillary tangle formation (see Fig. 2). A phase II clinical trial sponsored by the National Institute of Neurological Disorders and Stroke (NINDS) is currently evaluating the efficacy of lithium and/or divalproex in reducing phosphorylated tau in the CSF of AD patients.

References

1. Busser J, Geldmacher DS, Herrup K. Ectopic cell cycle proteins predict the site of neuronal cell death in Alzheimer's disease brain. J Neurosci 1998; 18:2801-2807.

2. Herrup K, Neve R, Ackerman S et al. Divide and die: Cell cycle events as triggers of nerve cell death. J Neurosci 2004; 24:9232-9239.

3. Hussemann JW, Nochlin D, Vincent I. Mitotic activation: A convergent mechanism for a cohort of neurodegenerative diseases. Neurobiol Aging 2000; 216:815-28.

4. Klein JA, Ackerman SL. Oxidative stress, cell cycle, and neurodegeneration. J Clin Invest 2003; 111:785-93.

5. Love S. Neuronal expression of cell cycle-related proteins after brain ischemia in man. Neurosci Lett 2003; 353:29-32.

6. McShea A, Harris PL, Webster KR et al. Abnormal expression of the cell cycle regulators p16 and CDK4 in Alzheimer's disease. Am J Pathol 1997; 150:1933-1939.

7. Ranganathan S, Bowser R. Alterations in G(1) to S phase cell cycle-regulators during amyotrophic lateral sclerosis. Am J Pathol 2003; 162:823-835.

8. Vincent I, Rosado M, Davis P. Mitotic mechanisms in Alzheimer's disease? J Cell Biol 1996; 132:413-425.

9. Heintz N. Cell death and the cell cycle: A relationship between transformation and neurodegeneration? Trends Biochem Sci 1993; 18:157-159.

10. Copani A, Condorelli F, Caruso A et al. Mitotic signaling by beta-amyloid causes neuronal death. FASEB J 1999; 13:2225-2234.

11. Copani A, Sortino MA, Caricasole A et al. Erratic expression of DNA polymerases by beta-amyloid causes neuronal death. FASEB J 2002; 16:2006-2008, (Epub 2002 Oct 18).

12. Grana X, Reddy EP. Cell cycle control in mammalian cells: Role of cyclins, cyclin dependent kinases (CDKs), growth suppressor genes and cyclin dependent kinase inhibitors (CKIs). Oncogene 1995; 11:211-219.

13. Freeman RF, Estus S, Johnson EM. Analysis of cell-related gene expression in post mitotic neurons: Selective induction of cyclin D1 during programmed cell death. Neuron 1994; 12:343-345.

14. Giovanni A, Wirtz-Brugger F, Keramaris E et al. Involvement of cell cycle elements, cyclin-dependent kinases, pRB, and E"F-DP, in β-amyloid-induced neuronal death. J Biol Chem 1999; 274:19011-19016.

15. Park DS, Levine B, Ferrari G et al. Cyclin-dependent kinase inhibitors and dominant-negative cyclin-dependent kinase 4 and 6 promote survival of NGF-deprived sympathetic neurons. J Neurosci 2004; 17:8975-8983.

16. Hubscher U, Nasheuer HP, Syvaoja JE. Eukaryotic DNA polymerases, a growing family. Trends Biochem Sci 2000; 25:143-147.

17. Friedberg EC, Wagner R, Radman M. Specialized DNA polymerases, cellular survival, and the genesis of mutations. Science 2002; 296:1627-1630.

18. Sobol R, Horton J, Kuhn R et al. Requirement of mammalian DNA polymerase-beta in base-excision repair. Nature 1996; 379:183-186.

19. Raji NS, Hari Krishna T, Rao KS. DNA-polymerase alpha, beta, delta and epsilon activities in isolated neuronal and astroglial cell fractions from developing and aging rat cerebral cortex. Int J Dev Neurosci 2002; 20:491-496.

20. Su JH, Deng G, Cotman CW. Neuronal DNA damage precedes tangle formation and is associated with up-regulation of nitrotyrosine in Alzheimer's disease brain. Brain Res 1997; 774:193-9.

21. Yang Y, Geldmacher DS, Herrup K. DNA replication precedes neuronal cell death in Alzheimer's disease. J Neurosci 2001; 21:2661-8.

22. Caricasole A, Copani A, Caraci F et al. Induction of dickkopf-1, a negative modulator of the Wnt pathway, is associated with neuronal degeneration in Alzheimer's brain. J Neurosci 2004; 24:6021-6027.

23. Zorn AM. Wnt signalling: Antagonistic Dickkopfs. Curr Biol 2001; 11:R592-5.

24. De Ferrari GV, Chacòn M, Barria MI et al. Activation of Wnt signaling rescues neurodegeneration and behavioural impairments induced by β-amyloid fibrils. Mol Psych 2003; 8:195-208.

25. Inestrosa N, De Ferrari GV, Garrido JL et al. Wnt signaling involvement in beta-amyloid-dependent neurodegeneration. Neurochem Int 2002; 41:341-4.

26. Willert K, Nusse R. Beta-catenin: A key mediator of Wnt signaling. Curr Opin Genet Dev 1998; 8:95-102.

27. Fuentalba RA, Farias G, Scheu J et al. Signal transduction during amyloid-β-peptide neurotoxicity: Role in Alzheimer disease. Brain Res Rev 2004; 47:275-289.

28. Liolitsa D, Powell J, Lovestone S. Genetic variability in the insulin signalling pathway may contribute to the risk of late onset Alzheimer's disease. J Neurol Neurosurg Psychiatry 2002; 73:261-6.

29. Sortino MA, Chisari M, Merlo S et al. Glia mediates the neuroprotective action of estradiol on β-amyloid-induced neuronal death. Endocrinology 2004; 145:5080-5086.

30. Bruno V, Sureda FX, Storto M et al. The neuroprotective activity of group-II metabotropic glutamate receptors requires new protein synthesis and involves a glial-neuronal signaling. J Neurosci 1997; 17:1891-1897.

31. Bruno V, Battaglia G, Casabona G et al. Neuroprotection by glial metabotropic glutamate receptors is mediated by transforming growth factor-beta. J Neurosci 1998; 18:9594-9600.

32. D'Onofrio M, Cuomo L, Battaglia G et al. Neuroprotection mediated by glial group-II metabotropic glutamate receptors requires the activation of the MAP kinase and the phosphatidylinositol-3-kinase pathways. J Neurochem 2001; 78:435-445.

33. Copani A, Bruno V, Battaglia G et al. Activation of metabotropic glutamate receptors protects cultured neurons against apoptosis induced by beta-amyloid peptide. Mol Pharmacol 1995; 47:890-897.

34. Alvarez G, Munoz-Montano JR, Satrustegui J et al. Regulation of tau phosphorylation and protection against beta-amyloid-induced neurodegeneration by lithium. Possible implications for Alzheimer's disease. Bipolar Disord 2002; 4:153-65.

Index